Methods of Fundamental Solutions in Solid Mechanics

Hui Wang
Qing-Hua Qin

ELSEVIER

Library of Congress Cataloging-in-Publication Data
A catalog record for this book is available from the Library of Congress

British Library Cataloguing-in-Publication Data
A catalogue record for this book is available from the British Library

ISBN: 978-0-12-818283-3

For information on all Elsevier publications visit our
website at https://www.elsevier.com/books-and-journals

Publisher: Matthew Deans
Acquisition Editor: Glyn Jones
Editorial Project Manager: Naomi Robertson
Production Project Manager: Sruthi Satheesh
Cover Designer: Mark Rogers

Typeset by TNQ Technologies

Working together
to grow libraries in
developing countries

www.elsevier.com • www.bookaid.org

Methods of Fundamental Solutions in Solid Mechanics

To our parents and families.

Contents

Part II
Applications of the meshless method

4. Meshless analysis for thin beam bending problems

5. Meshless analysis for thin plate bending problems

6. Meshless analysis for two-dimensional elastic problems

7. Meshless analysis for plane piezoelectric problems

About the authors

Dr. Hui Wang was born in Mengjin County, China in 1976. He received his bachelor's degree in Theoretical and Applied Mechanics from Lanzhou University, China, in 1999. Subsequently he joined the College of Science as an assistant lecturer at Zhongyuan University of Technology (ZYUT) and spent 2 years teaching at ZYUT. He earned his master's degree from Dalian University of Technology in 2004 and doctoral degree from Tianjin University in 2007, both of which are in Solid Mechanics. Since 2007, he has worked at College of Civil Engineering and Architecture, Henan University of Technology, as a lecturer. He was promoted to Associate Professor in 2009 and Professor in 2015. From August 2014 to August 2015, he worked at the Australian National University (ANU) as a visiting scholar, and then from February 2016 to February 2017, he joined the ANU as a Research Fellow.

His research interests include computational mechanics, meshless methods, hybrid finite element methods, and the mechanics of composites. So far, he has authored three academic books by CRC Press and Tsinghua University Press, respectively, 7 book chapters, and 62 academic journal papers (47 indexed by SCI and 8 indexed by EI). In 2010, He was awarded the Australia Endeavor Award.

Dr. Qing-Hua Qin was born in Yongfu County, China in 1958. He received his Bachelor of Engineering degree in mechanical engineering from Chang An University, China, in 1982, earning his Master of Science degree in 1984 and doctoral degree in 1990 from Huazhong University of Science and Technology (HUST) in China. Both degrees are in applied mechanics. He joined the Department of Mechanics as an associate lecturer at HUST in 1984 and was promoted to lecturer in mechanics in 1987 during his Ph.D. candidature period. After spending 10 years lecturing at HUST, he was awarded the DAAD/K.C. Wong research fellowship in 1994, which enabled him to work at the University of Stuttgart in Germany for 9 months. In 1995 he left HUST to take up a postdoctoral research fellowship at Tsinghua University, China, where he worked until 1997. He was awarded a Queen Elizabeth II fellowship in 1997 and a Professorial fellowship in 2002 at the University of Sydney (where he stayed there until December 2003), both by the Australian Research Council, and he is currently working as a professor in the Research

School of Electrical, Energy and Materials Engineering at the Australian National University, Australia. He was appointed a guest professor at HUST in 2000. He has published more than 300 journal papers and 7 books in the field of applied mechanics.

Preface

Since the basic concept behind the method of fundamental solutions (MFS) was developed primarily by V. D. Kupradze and M. A. Alexidze in 1964, the meshless MFS has become an effective tool for the solution of a large variety of physical and engineering problems, such as potential problems, elastic problems, crack problems, fluid problems, piezoelectric problems, antiplane problems, inverse problems, and free-boundary problems. More recently, it has been extended to deal with inhomogeneous partial differential equations, partial differential equations with variable coefficients, and time-dependent problems, by introducing radial basis function interpolation (RBF) for particular solutions caused by inhomogeneous terms.

The great advantage of the MFS over other numerical methods is that it can easily be implemented for physical and engineering problems in two- or three-dimensional regular and irregular domains. This means that it can be programmed with simple codes by users and no additional professional skills are required. Clearly, there are still some limitations in the range of implementation of the MFS, as pointed out in Chapter 1 of this book. However, the MFS has sufficiently shown its success as an executable numerical technique in a simple form via various engineering applications. Hence, it is warranted to present some of the recent significant developments in the MFS for the further understanding of the physical and mathematical characteristics of the procedures of MFS which is the main objective of this book.

This book covers the fundamentals of continuum mechanics, the basic concepts of the MFS, and its methodologies and extensive applications to various engineering inhomogeneous problems. This book consists of eight chapters. In Chapter 1, current meshless methods are reviewed, and the advantages and disadvantages of MFS are stressed. In Chapter 2, the basic knowledge involved in this book is provided to give a complete description. In Chapter 3, the basic concepts of fundamental solutions and RBF are presented. Starting from Chapter 4, some engineering problems including Euler—Bernoulli beam bending, thin plate bending, plane elasticity, piezoelectricity, and heat conduction are solved in turn by combining the MFS and the RBF interpolation.

Hui Wang
Qing-Hua Qin

Acknowledgments

The motivation that led to the development of this book based on our extensive research since 2001, in the context of the method of fundamental solutions (MFS). Some of the research results presented in this book were obtained by the authors at the Department of Engineering Mechanics of Henan University of Technology, the Research School of Engineering of the Australian National University, the Department of Mechanics of Tianjin University, and the Department of Mechanics of Dalian University of Technology. Thus, support from these universities is gratefully acknowledged.

Additionally, many people have been most generous in their support of this writing effort. We would like especially to thank Professor Xingpei Liang of Henan University of Technology for his meaningful discussions. Special thanks go to Senior Editors Jianbo Liu and Chao Wang, Senior Copyright Manager Xueying Zou of Higher Education Press as well as Glyn Jones, Naomi Robertson and Sruthi Satheesh of Elsevier Inc. for their commitment to the publication of this book. Finally, we are very grateful to the reviewers who made suggestions and comments for improving the quality of the book.

List of abbreviations

AEM Analog equation method
BEM Boundary element method
BVP Boundary value problem
DOF Degrees of freedom
FE Finite element
FEM Finite element method
FGM Functionally graded materials
FS Fundamental solution
MFS Method of fundamental solutions
MQ Multiquadric
PDE Partial differential equation
PS Power spline
PZT Lead zirconate titanate
RBF Radial basis function
TPS Thin plate spline

Part I

Fundamentals of meshless methods

Chapter 1

Overview of meshless methods

Chapter outline

1.1 Why we need meshless methods

Initial and/or boundary value problems such as heat transfer, wave propagation, elasticity, fluid flow, piezoelectricity, and electromagnetics can be found in almost every field of science and engineering applications and are generally modeled by partial differential equations (PDEs) with prescribed boundary conditions and/or initial conditions in a given domain. Thus PDEs are fundamental to the modeling of physical problems. Consequently, clear understanding of the physical meaning of solutions of these equations has become very important to engineers and mathematicians, so that the solutions

Methods of Fundamental Solutions in Solid Mechanics. https://doi.org/10.1016/B978-0-12-818283-3.00001-4
© 2019 Higher Education Press. Published by Elsevier Inc. All rights reserved.

can be used to explain natural phenomena or to design and develop new structures and materials.

To this end, theoretical and experimental methods were first developed. However, PDEs usually involve higher-order partial differentials to spatial or time variables, but theoretical analysis is operable only for problems with simple expressions of PDEs, simple domain shapes, and simple boundary conditions. Most engineering problems cannot be solved analytically. Besides, due to limitations to experimental conditions, most engineering problems cannot be tested experimentally, and even if the problem can be tested, the process is time-consuming and expensive; moreover, not all of the relevant information can be measured experimentally. Hence, ways to find good approximate solutions using numerical methods should be very interesting and helpful.

Currently, solutions of ordinary/boundary value problems can be computed directly or iteratively using numerical methods, such as the finite element method (FEM) [1−3], the hybrid finite element method (HFEM) [4−8], the boundary element method (BEM) [9−12], the finite difference/volume method (FDM/FVM) [2,13], and the meshless method [14,15], etc.

Among these methods, conventional element-dependent methods such as the FEM, the HFEM, the BEM, and the FVM require domain or boundary element partition to model complex multiphysics problems. For example, in the classic FEM, which was originally developed by Richard Courant in 1943 for structural torsion analysis [16], a continuum with a complex boundary shape can geometrically be divided into a finite number of elements, generally termed finite elements. These individual finite elements are connected through nodes to form a topological mesh. In each individual element, the proper interpolating function is chosen to approximate the real distribution of the physical field. Then all elements are integrated into a weak-form integral functional to produce the solving system of equations, so that the nodal field as the primary variable in the final solving system can be fully determined. Due to such features as versatility of meshing for complex geometry and flexibility of solving strategy for complex problems, the FEM has become one of the most robust and well-developed procedures for boundary/ordinary value problems, and it has been successfully applied in many engineering applications [1].

There are, however, some inherent limitations for use of the FEM. Firstly, mesh generation is usually time-consuming in the FEM, and one must spend much time and effort for complex data management and treatment. Secondly, mesh generation involves complex algorithms, and this is not always possible for problems with complex domains, especially three-dimensional geometrically complex domains. In some cases, users must make partitions to create relatively regular subdomains for applying mesh generation. Thirdly, as a displacement-based numerical method, the FEM can relatively accurately predict displacement, rather than stress that involves higher-order derivatives of displacement to spatial variables. Moreover, stresses obtained in the FEM are often discontinuous across the common interface of adjacent elements, because of the element-wise continuous nature of the displacement field assumed in the FEM formulation. Special smoothed treatment techniques are required in the postprocessing stage to recover stresses. Fourthly, numerical accuracy in the FEM is significantly dependent on mesh density, so a remeshing operation is usually required to achieve a desired accuracy. Multiple meshing trials may be needed for that purpose. More importantly, for large deformation problems, adaptive mesh is required to avoid element distortion during deformation. The remeshing process at each step often leads to additional computational time as well as degradation of accuracy in the solution. Finally, it is difficult for the FEM to effectively simulate crack growth and material fracture because the continuous elements in the FEM cannot easily model discontinuous fields. All these limitations can be attributed to the use of elements or mesh in the FEM formulation.

As an alternative to the FEM, the element-dependent BEM attempts to find numerical solutions of boundary integral equations by incorporating element discretization over the domain boundary [2,10,17,18]. In the BEM, the boundary integral equation may be regarded as an exact solution of the governing partial differential equation of the problem, to which the fundamental solution should be explicitly available, and it can be derived by introducing the physical definition of fundamental solution to include boundary values in the representation formula. Such distinguishing features give the BEM the following advantages over the FEM: (1) Only the boundary of the domain needs to be discretized; as a result, very simple data input and storage are involved. (2) The solution in the interior of the domain is approximated with a rather high convergence rate and accuracy. (3) Exterior problems with unbounded domains are more properly solved by the BEM, because the remote condition can be exactly satisfied. However, the BEM also experiences some difficulties. For example, the BEM is greatly dependent of the fundamental solution of the partial differential equation. So, problems with inhomogeneities or nonlinear differential equations are in general not easily handled by pure BEM. Additionally, the BEM requires more mathematical foundation, compared to the FEM, to obtain the boundary integral equation and evaluate boundary integrals, because the kernels in boundary integrals are in general singular. All these disadvantages limit BEM applications, so sometimes coupling of other methods must be implemented [19].

In contrast, meshless methods based on boundary and/or interior collocations in a domain can remove limitations such as in the FEM and BEM, significantly reducing the effort of model discretization preparation. To illustrate this, Fig. 1.1 displays domain discretization by elements and collocations in the conventional FEM and the meshless method. In general, the term "meshless method" refers to the class of numerical techniques that do not require any predefined mesh information for domain discretization, relying simply on either global or localized interpolation on nonordered spatial points scattered within the problem domain as well as on its boundaries.

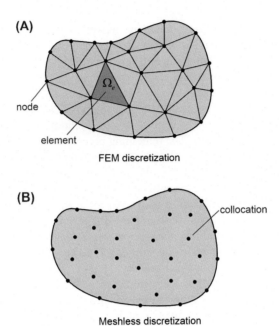

(A)

node

element

FEM discretization

(B)

collocation

Meshless discretization

FIGURE 1.1 FEM and meshless discretization in a 2D domain.

These scattered points are not geometrically connected to form a mesh. Thus meshless methods present great potential for engineering applications.

1.2 Review of meshless methods

Meshless methods are used to solve PDE in strong or weak form by arbitrarily distributed collocations in the solution domain, and these points contribute to the approximation by assumed global or local basis functions.

As in the classification of FEM and BEM, meshless methods may also be divided into two categories: domain-type meshless methods and boundary-type meshless methods. Domain-type meshless methods, as the name implies, employ arbitrarily distributed interior and boundary collocations to represent the problem domain and its boundary. Then the field variable (e.g., displacement, temperature, pressure) at any field point within the problem domain is interpolated using function values at a local or global set of nodes around the field point. Subsequently, a weak-form integral equation such as the Galerkin integral equation or a strong-form collocation equation is introduced to include the assumed field variables in producing a system of equations associated with the variables. Strong-form domain-type meshless methods, include smoothed particle hydrodynamics (SPH) [20], the finite point method (FPM) [21], the point collocation method (PCM) [22], the Kansa collocation method [23], and the least-squares collocation meshless method [24], etc. Weak-form domain-type meshless methods include the diffuse element method (DEM) [25], the element-free Galerkin method (EFGM) [26], the reproducing kernel particle method (RKPM) [27], partition of unity finite element methods (PUFEM) [28], the Hp-meshless clouds method [29], the meshless local Petrov-Galerkin method (MLPG) [30], and the point interpolation method (PIM) [31], etc. In contrast to the strong-from meshless methods, the meshless methods based on the weak form are more robust and steady. However, the latter generally involve more complex formulations.

Unlike domain-type meshless methods, boundary-type meshless methods require only boundary collocation in the computation. In that case, the T-complete function basis [5,6] or the fundamental solution basis [10] is typically employed to approximate the field variable, so it can exactly satisfy the governing equations of the problem. For example, Zhang et al. presented the solution procedure of the hybrid boundary node method (BNM) [32] that combines the modified variational functional and the moving least-square (MLS) approximation based only on a group of arbitrarily distributed boundary nodes. This method is a weak-form method and only involves integration computation over the boundary. However, MLS interpolations lack the delta function property, so it is difficult to satisfy the boundary conditions accurately in the BNM. In an improvement over the BNM, the boundary point interpolation method (BPIM) [33] employs shape functions having the Kronecker delta function property, so boundary conditions can be enforced as easily as in the conventional BEM.

Besides, some pure boundary collocation methods exist, including the method of fundamental solutions (MFS) [14,34], boundary particle method (BPM) [35,36], and Trefftz boundary collocation method (TBCM) [37,38]. Among these, the MFS is the most powerful and popular, when fundamental solutions of a problem are explicitly defined. It was initially introduced by Kupradze and Aleksidze in 1964 [39]. Over the past 30 years, MFS has been widely used for the numerical approximation of a large variety of physical problems [34]. The general concept of the MFS is that the solution (field variable) approximated by a linear combination of fundamental solutions with respect to source points located outside the solution domain can exactly satisfy the governing partial differential equation of the problem, so only boundary conditions at boundary collocations need to be satisfied. The MFS inherits all the advantages of the BEM [10]; for example, it does not require discretization over the domain, unlike domain mesh discretization methods such as the FEM. Moreover, unlike the BEM, integrations over the domain boundary are avoided. The solution in the interior of the domain is evaluated also without any additional integration equation. Furthermore, the derivatives of the solution are calculated directly from the MFS representation. More importantly, implementation of the MFS is very easy and direct, even for problems in three-dimensional and irregular domains, and only a little data preparation is required. Besides, high numerical accuracy and fast convergence has been proven [40–42]. Thus, it is employed and discussed in this book.

1.3 Basic ideas of the method of fundamental solutions

To illustrate the basic concept of the MFS, the following two-dimensional homogeneous partial differential equation is considered:

$$\mathbf{L}u(\mathbf{x}) = 0, \quad \mathbf{x} \in \Omega \tag{1.1}$$

where $\mathbf{L}$ is a partial differential operator, $\mathbf{x} = (x, y)$ is an arbitrary point in a bounded domain $\Omega \subset R^2$, R^2 denotes the Euclidean 2-space, and u is the unknown field variable. It is noted that in the book we sometimes write $\mathbf{x} = (x_1, x_2)$ rather than $\mathbf{x} = (x, y)$ for the sake of convenience.

For a complete boundary value problem, the proper boundary conditions should be complemented, for example,

$$u(\mathbf{x}) = \bar{u}, \quad \mathbf{x} \in \Gamma_1$$
$$\frac{\partial u(\mathbf{x})}{\partial n} = \bar{q}, \quad \mathbf{x} \in \Gamma_2 \tag{1.2}$$

where $\Gamma_1 \cup \Gamma_2 = \partial\Omega$, $\Gamma_1 \cap \Gamma_2 = \varnothing$, $\bar{u}$ and $\bar{q}$ are given functions, and n is the unit normal to the boundary $\partial\Omega$, as indicated in Fig. 1.2.

1.3.1 Weighted residual method

The boundary value problem (BVP) described in the governing Eq. (1.1) and the boundary conditions (1.2) can be solved in an approximate manner, by which the field variable u is approximated in a general form by

$$u(\mathbf{x}) \approx u^h(\mathbf{x}) = \sum_{i=1}^{m} \alpha_i f_i(\mathbf{x}) \tag{1.3}$$

where $f_i(\mathbf{x})$ is the ith basis function term, α_i is the unknown coefficient for the ith basis function term, and m is the number of basis functions.

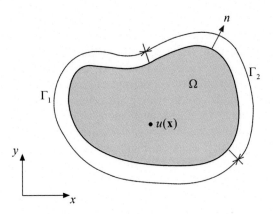

FIGURE 1.2   Schematic of boundary value problem.

In practice, the approximation function given in Eq. (1.3) cannot exactly satisfy the governing equation and the boundary conditions, due to the truncated approximation. Hence, substituting Eq. (1.3) into Eqs. (1.1) and (1.2), we usually have

$$\mathbf{L}u^h(\mathbf{x}) \neq 0, \quad \mathbf{x} \in \Omega \tag{1.4}$$

and

$$u^h(\mathbf{x}) \neq \bar{u}, \quad \mathbf{x} \in \Gamma_1$$
$$\frac{\partial u^h(\mathbf{x})}{\partial n} \neq \bar{q}, \quad \mathbf{x} \in \Gamma_2 \tag{1.5}$$

If the following residual functions can be defined,

$$R_g(\mathbf{x}) = \mathbf{L}u^h(\mathbf{x}), \quad \mathbf{x} \in \Omega \tag{1.6}$$

$$R_{b1}(\mathbf{x}) = u^h(\mathbf{x}) - \bar{u}, \quad \mathbf{x} \in \Gamma_1$$
$$R_{b2}(\mathbf{x}) = \frac{\partial u^h(\mathbf{x})}{\partial n} - \bar{q}, \quad \mathbf{x} \in \Gamma_2 \tag{1.7}$$

then Eqs. (1.4) and (1.5) can be respectively rewritten as

$$R_g(\mathbf{x}) \neq 0, \quad \mathbf{x} \in \Omega \tag{1.8}$$

and

$$R_{b1}(\mathbf{x}) \neq 0, \quad \mathbf{x} \in \Gamma_1$$
$$R_{b2}(\mathbf{x}) \neq 0, \quad \mathbf{x} \in \Gamma_2 \tag{1.9}$$

Theoretically, the proper approximate function should make the residuals at each point $\mathbf{x}$ inside the domain or on the boundary as small as possible. However, such strong-form satisfaction is difficult to derive. In the practical operation, we usually force the residuals to zero in an averaged sense by setting the so-called weighted integrals of the residuals to zero, i.e.,

$$\int_\Omega w_i R_g(\mathbf{x}) d\Omega + \int_{\Gamma_1} v_i R_{b1}(\mathbf{x}) d\Gamma + \int_{\Gamma_2} s_i R_{b2}(\mathbf{x}) d\Gamma = 0 \tag{1.10}$$

where w_i, v_i, and s_i are a set of given weight functions corresponding to the ith point, and $i = 1, 2, \ldots, n$.

In the weighted residual method, the selection of weight functions determines the type and performance of numerical approximations. In this book, the collocation-type meshless method is discussed. In the collocation method, a set of points that are distributed in the domain are chosen for approximation, and we can try to satisfy the BVP at only these points, rather than satisfying the BVP in an integral form. To obtain the collocation solution, we select the standard Dirac delta function as the weight functions in Eq. (1.10), i.e.,

$$w_i = \delta(\mathbf{x}, \mathbf{x}_i)$$
$$v_i = \delta(\mathbf{x}, \mathbf{x}_i) \tag{1.11}$$
$$s_i = \delta(\mathbf{x}, \mathbf{x}_i)$$

where $\delta(\mathbf{x}, \mathbf{x}_i)$ is the Dirac delta function centered at the point $\mathbf{x}_i$ and has the following properties:

$$\delta(\mathbf{x}, \mathbf{x}_i) = \begin{cases} 0, & \mathbf{x} \neq \mathbf{x}_i \\ \infty, & \mathbf{x} = \mathbf{x}_i \end{cases} \tag{1.12}$$

and

$$\int_{-\infty}^{\infty} \delta(\mathbf{x}, \mathbf{x}_i) f(\mathbf{x}) d\Omega = f(\mathbf{x}_i) \tag{1.13}$$

Thus, the collocation method can be derived from the weighted residual formulation by substituting Eq. (1.11) into it:

$$\int_{\Omega} \delta(\mathbf{x}, \mathbf{x}_i) R_g(\mathbf{x}) d\Omega + \int_{\Gamma_1} \delta(\mathbf{x}, \mathbf{x}_i) R_{b1}(\mathbf{x}) d\Gamma + \int_{\Gamma_2} \delta(\mathbf{x}, \mathbf{x}_i) R_{b2}(\mathbf{x}) d\Gamma = 0 \tag{1.14}$$

which results in

$$R_g(\mathbf{x}_i) + R_{b1}(\mathbf{x}_i) + R_{b2}(\mathbf{x}_i) = 0 \tag{1.15}$$

Eq. (1.15) shows that the residuals are zero at n points chosen in the problem domain. Specially, if the approximation can exactly satisfy the governing Eq. (1.1), i.e., the Trefftz approximation and fundamental solution approximation taken into account in this book, the domain residual R_g should be zero. In that case, Eq. (1.15) can be reduced to

$$R_{b1}(\mathbf{x}_i) + R_{b2}(\mathbf{x}_i) = 0, \quad \mathbf{x}_i \in \Gamma \tag{1.16}$$

which means that only boundary collocations can be chosen to force the boundary residuals to zero at those collocations. This can be regarded as the theoretical basis of the MFS described next.

1.3.2 Method of fundamental solutions

In the application of the MFS, the solution of problems governed by Eqs. (1.1) and (1.2) is approximated as a linear combination of fundamental solutions:

$$u(\mathbf{x}) \approx \sum_{j=1}^{N} \alpha_j G^*(\mathbf{x}, \mathbf{x}_{sj}) = \mathbf{G}\boldsymbol{\alpha}, \quad \mathbf{x} \in \Omega, \quad \mathbf{x}_{sj} \notin \Omega \tag{1.17}$$

where N is the number of source points $\{\mathbf{x}_{sj}\}_{j=1}^{N}$ located outside the domain, $G^*(\mathbf{x}, \mathbf{x}_{sj})$ centered at $\mathbf{x}_{sj}$ is the fundamental solution of the partial differential Eq. (1.1) in the spatial coordinate $\mathbf{x} = (x, y)$ [43,44], α_j is the coefficient for $G^*(\mathbf{x}, \mathbf{x}_{sj})$ that is yet to be determined, and

$$\mathbf{G} = [G^*(\mathbf{x}, \mathbf{x}_{s1}) \quad \dots \quad G^*(\mathbf{x}, \mathbf{x}_{sN})] \tag{1.18}$$

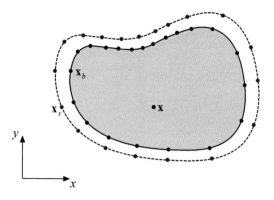

FIGURE 1.3 Collocations in the MFS.

$$\boldsymbol{\alpha} = \begin{bmatrix} \alpha_1 \\ \vdots \\ \alpha_N \end{bmatrix} \tag{1.19}$$

Obviously, the approximation (1.17) exactly satisfies the partial differential Eq. (1.1) due to the nonoverlapping of the source points $\mathbf{x}_{sj}$ ($j = 1, 2, ..., N$) and the field point $\mathbf{x}$. Therefore, the approximation (1.17) just needs to satisfy the specified boundary conditions (1.2). To determine all known coefficients in Eq. (1.17), we collocate the approximate solution (1.17) on a finite set of boundary points $\{\mathbf{x}_{b1}, \mathbf{x}_{b2}, ..., \mathbf{x}_{bM}\} \in \Gamma$, as shown in Fig. 1.3. M is the total number of boundary collocations. If M_1 and M_2 represent the number of collocations on the boundaries Γ_1 and Γ_2, respectively, and $M = M_1 + M_2$, we have

$$\begin{cases} \displaystyle\sum_{j=1}^{N} G^*(\mathbf{x}_{bi}, \mathbf{x}_{sj})\alpha_j = \bar{u}(\mathbf{x}_{bi}), & i = 1, ..., M_1 \\ \displaystyle\sum_{j=1}^{N} \frac{\partial G^*(\mathbf{x}_{bk}, \mathbf{x}_{sj})}{\partial n}\alpha_j = \bar{q}(\mathbf{x}_{bk}), & k = M_1 + 1, ..., M \end{cases} \tag{1.20}$$

where $\mathbf{x}_{bi}$ and $\mathbf{x}_{bk}$ are collocations on the boundary segment Γ_1 and Γ_2, respectively.

Eq. (1.20) can be rewritten in the following matrix form:

$$\begin{bmatrix} H_{11} & \cdots & H_{1N} \\ \vdots & \ddots & \vdots \\ H_{M_1 1} & \cdots & H_{M_1 N} \\ H_{(M_1+1)1} & \cdots & H_{(M_1+1)N} \\ \vdots & \ddots & \vdots \\ H_{M1} & \cdots & H_{MN} \end{bmatrix} \begin{bmatrix} \alpha_1 \\ \alpha_2 \\ \vdots \\ \alpha_N \end{bmatrix} = \begin{bmatrix} b_1 \\ \vdots \\ b_{M_1} \\ b_{M_1+1} \\ \vdots \\ b_M \end{bmatrix} \tag{1.21}$$

where

$$H_{ij} = \begin{cases} G^*(\mathbf{x}_{bi}, \mathbf{x}_{sj}), & i = 1, ..., M_1, \quad j = 1, 2, ..., N \\ \dfrac{\partial G^*(\mathbf{x}_{bk}, \mathbf{x}_{sj})}{\partial n}, & k = M_1 + 1, ..., M, \quad j = 1, 2, ..., N \end{cases} \tag{1.22}$$

and

$$b_i = \begin{cases} \bar{u}(\mathbf{x}_{bi}), & i = 1, ..., M_1 \\ \bar{q}(\mathbf{x}_{bk}), & k = M_1 + 1, ..., M \end{cases} \tag{1.23}$$

Explicitly, this leads to an $M \times N$-dimensional collocation linear system of equations that can generally be written as

$$\sum_{j=1}^{N} H_{ij}\alpha_j = b_i, \quad i = 1, 2, ...M \tag{1.24}$$

or in matrix form

$$\mathbf{H}_{M \times N}\boldsymbol{\alpha}_{N \times 1} = \mathbf{b}_{M \times 1} \tag{1.25}$$

where $\mathbf{H}$ is the $M \times N$-dimensional collocation matrix with elements H_{ij}, $\boldsymbol{\alpha}$ is the $N \times 1$ vector of unknowns, and $\mathbf{b}$ is the $M \times 1$ known vector.

To obtain a unique solution for the linear system of Eq. (1.25), we usually require $M \geq N$. If $M = N$, the linear system of Eq. (1.25) can be solved directly using the Gaussian elimination technique or a regularization technique such as the Tichonov regularization or the truncated singular value decomposition if the matrix $\mathbf{H}$ is ill-conditioned, whereas if $M > N$ we must employ the linear least-square technique to replace the rectangular $M \times N$ overdetermined system of Eq. (1.25) with a square $N \times N$ determined system. For example, minimization of the Euclidean error norm of Eq. (1.25)

$$\Pi = \sum_{i=1}^{M} \left(\sum_{j=1}^{N} H_{ij}\alpha_j - b_i \right)^2 \tag{1.26}$$

yields

$$\frac{\partial \Pi(\boldsymbol{\alpha})}{\partial \alpha_j} = 0, \quad j = 1, 2, ..., N \tag{1.27}$$

from which we obtain

$$\sum_{i=1}^{M} \left(\sum_{j=1}^{N} H_{ij}\alpha_j - b_i \right) H_{ik} = 0, \quad k = 1, 2, ...N \tag{1.28}$$

Eq. (1.28) can be rewritten in matrix form as

$$(\mathbf{H}^{\mathrm{T}}\mathbf{H})_{N \times N} \boldsymbol{\alpha}_{N \times 1} = (\mathbf{H}^{\mathrm{T}}\mathbf{b})_{N \times 1} \tag{1.29}$$

where the superscript "T" denotes the transpose of a matrix.

Solving Eq. (1.29) for the unknown coefficient vector $\boldsymbol{\alpha}$, we obtain

$$\boldsymbol{\alpha} = (\mathbf{H}^{\mathrm{T}}\mathbf{H})^{-1}(\mathbf{H}^{\mathrm{T}}\mathbf{b}) \tag{1.30}$$

and then the field variable at arbitrary point $x \in \Omega$ can be evaluated by Eq. (1.17).

One of the obvious advantages of such a method is the fact that there is no need for any mesh generation at any function evaluation, and only boundary collocations are required. This means that the computation is economical in terms of time. Other advantages of the MFS include its simple formulation and high precision that can be illustrated by its application to two-dimensional Laplace problems.

1.4 Application to the two-dimensional Laplace problem

To illustrate the MFS solution procedure, we consider a Laplace problem given in a two-dimensional Cartesian coordinate domain. The related source code is provided to better understand the MFS implementation.

1.4.1 Problem description

We consider a bounded simply or multiply connected region Ω in the plane xy. For the two-dimensional Laplace problem defined in the domain Ω, the governing partial differential equation related to the potential function u can be written as

$$\nabla^2 u(\mathbf{x}) = 0, \quad \mathbf{x} \in \Omega \tag{1.31}$$

where

$$\nabla^2 = \frac{\partial^2}{\partial x^2} + \frac{\partial^2}{\partial y^2} \tag{1.32}$$

represents the two-dimensional Laplace operator, and $\mathbf{x} = (x, y)$ is the arbitrary point in the plane.

Moreover, the governing Eq. (1.31) should be complemented by the following boundary conditions to form a well-defined BVP:

$$\begin{aligned} u &= f \quad \text{on } \Gamma_1 \\ q &= g \quad \text{on } \Gamma_2 \end{aligned} \tag{1.33}$$

where $q = \partial u / \partial n$ denotes the outward normal derivative of u on the boundary $\partial\Omega$, and f and g are given values defined on Γ_1 and Γ_2, respectively. Here, $\partial\Omega = \Gamma_1 \cup \Gamma_2$ and $\Gamma_1 \cap \Gamma_2 = \varnothing$.

The MFS employs the fundamental solution of a problem to approximate the unknown potential field u. Therefore, the fundamental solution of the problem should be explicitly formulated. Here, the fundamental solution to the two-dimensional Laplace operator in Eq. (1.31) can be given by

$$u^*(\mathbf{x}, \mathbf{x}_s) = -\frac{1}{2\pi} \ln r \tag{1.34}$$

where

$$r = \|\mathbf{x} - \mathbf{x}_s\| \tag{1.35}$$

is the Euclidean distance, $\mathbf{x}_s$ is a source point at which a unit source intensity is applied, and $\mathbf{x}$ is an arbitrary field point in an infinite two-dimensional domain.

Correspondingly, the first-order derivatives of u^* can be obtained as

$$\begin{aligned}
\frac{\partial u^*(\mathbf{x}, \mathbf{x}_s)}{\partial x} &= -\frac{1}{2\pi r} \frac{\partial r}{\partial x} = -\frac{x - x_s}{2\pi r^2} \\
\frac{\partial u^*(\mathbf{x}, \mathbf{x}_s)}{\partial y} &= -\frac{1}{2\pi r} \frac{\partial r}{\partial y} = -\frac{y - y_s}{2\pi r^2}
\end{aligned} \tag{1.36}$$

Figs. 1.4−1.6 plot the variations of u^* and its derivatives for the source point $\mathbf{x}_s = (0, 0)$, respectively, from which it is observed that the fundamental solution u^* and its derivatives $\partial u^*/\partial x$ and $\partial u^*/\partial y$ show singularity at the source point. Fortunately, the source point cannot overlap with the field point in the theory of MFS, because it is arranged outside the domain of interest.

1.4.2 MFS formulation

Based on the fundamental solution of the problem given in Eq. (1.34), the following scheme for the implementation of the MFS can be obtained. Choosing the set of source points $\{\mathbf{x}_{sk}\}_{k=1}^{N}$ outside the computing domain Ω,

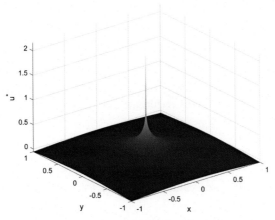

FIGURE 1.4 Variation of u^* for the two-dimensional Laplace operator at $\mathbf{x}_s = (0, 0)$.

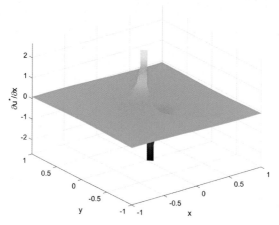

FIGURE 1.5 Variation of $\partial u^*/\partial x$ for the two-dimensional Laplace operator at $\mathbf{x}_s = (0, 0)$.

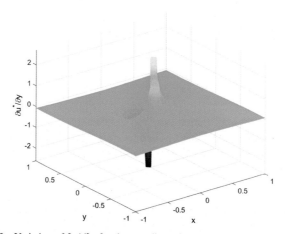

FIGURE 1.6 Variation of $\partial u^*/\partial y$ for the two-dimensional Laplace operator at $\mathbf{x}_s = (0, 0)$.

we can construct the approximation for the solution u of Eqs. (1.31) and (1.33) as follows:

$$u(\mathbf{x}) = \sum_{k=1}^{N} \alpha_k u^*(\mathbf{x}, \mathbf{x}_{sk}) \tag{1.37}$$

or in matrix form:

$$u = \mathbf{U}\boldsymbol{\alpha} \tag{1.38}$$

with

$$\mathbf{U} = [u^*(\mathbf{x}, \mathbf{x}_{s1}) \quad u^*(\mathbf{x}, \mathbf{x}_{s2}) \quad \dots \quad u^*(\mathbf{x}, \mathbf{x}_{sN})]_{1 \times N} \tag{1.39}$$

$$\boldsymbol{\alpha}^{\mathrm{T}} = [\alpha_1 \quad \alpha_2 \quad \cdots \quad \alpha_N]_{1 \times N} \tag{1.40}$$

Further, the first-order derivatives of u can be approximated as

$$q_x = \frac{\partial u}{\partial x} = \sum_{k=1}^{N} \alpha_k \frac{\partial u^*(\mathbf{x}, \mathbf{x}_{sk})}{\partial x} = \frac{\partial \mathbf{U}}{\partial x} \boldsymbol{\alpha}$$

$$q_y = \frac{\partial u}{\partial y} = \sum_{k=1}^{N} \alpha_k \frac{\partial u^*(\mathbf{x}, \mathbf{x}_{sk})}{\partial y} = \frac{\partial \mathbf{U}}{\partial y} \boldsymbol{\alpha} \tag{1.41}$$

Then, the outward normal derivative of u on Γ_2 can be expressed as

$$q = \frac{\partial u}{\partial n} = q_x n_x + q_y n_y = \sum_{k=1}^{N} \alpha_k \left(\frac{\partial u^*(\mathbf{x}, \mathbf{x}_{sk})}{\partial x} n_x + \frac{\partial u^*(\mathbf{x}, \mathbf{x}_{sk})}{\partial y} n_y \right)$$

$$= \sum_{k=1}^{N} \alpha_k q^*(\mathbf{x}, \mathbf{x}_{sk}) = \mathbf{Q}\boldsymbol{\alpha} \tag{1.42}$$

where n_x and n_y are components of the outward normal vector $\mathbf{n}$, and

$$q^*(\mathbf{x}, \mathbf{x}_{sk}) = \frac{\partial u^*(\mathbf{x}, \mathbf{x}_{sk})}{\partial x} n_x + \frac{\partial u^*(\mathbf{x}, \mathbf{x}_{sk})}{\partial y} n_y \tag{1.43}$$

$$\mathbf{Q} = [q^*(\mathbf{x}, \mathbf{x}_{s1}) \quad q^*(\mathbf{x}, \mathbf{x}_{s2}) \quad \cdots \quad q^*(\mathbf{x}, \mathbf{x}_{sN})]_{1 \times N} \tag{1.44}$$

In the preceding approximations, the coefficients $\{\alpha_k\}_{k=1}^{N}$ can be determined by the satisfaction of these boundary conditions at collocation points on the boundary $\partial\Omega$; that is, if we take N collocations $\{\mathbf{x}_k\}_{k=1}^{N}$ on $\partial\Omega$ and then impose the given boundary conditions, we have

$$u(\mathbf{x}_k) = \sum_{k=1}^{N} \alpha_k u^*(\mathbf{x}_k, \mathbf{x}_{sk}) = f(\mathbf{x}_k), \quad \mathbf{x}_k \in \Gamma_1, \ k = 1, 2, \dots, N_1$$

$$q(\mathbf{x}_k) = \sum_{k=1}^{N} \alpha_k q^*(\mathbf{x}_k, \mathbf{x}_{sk}) = g(\mathbf{x}_k), \quad \mathbf{x}_k \in \Gamma_2, \ k = 1, 2, \dots, N_2 \tag{1.45}$$

where N_1 and N_2 are the number of collocations on Γ_1 and Γ_2, respectively, and $N = N_1 + N_2$.

Eq. (1.45) can be rewritten in matrix notation as

$$\begin{bmatrix} u^*(\mathbf{x}_1, \mathbf{x}_{s1}) & u^*(\mathbf{x}_1, \mathbf{x}_{s2}) & \cdots & u^*(\mathbf{x}_1, \mathbf{x}_{sN}) \\ \vdots & \vdots & \vdots & \vdots \\ u^*(\mathbf{x}_{N_1}, \mathbf{x}_{s1}) & u^*(\mathbf{x}_{N_1}, \mathbf{x}_{s2}) & \cdots & u^*(\mathbf{x}_{N_1}, \mathbf{x}_{sN}) \\ \hline q^*(\mathbf{x}_1, \mathbf{x}_{s1}) & q^*(\mathbf{x}_1, \mathbf{x}_{s2}) & \cdots & q^*(\mathbf{x}_1, \mathbf{x}_{s2}) \\ \vdots & \vdots & \vdots & \vdots \\ q^*(\mathbf{x}_{N_1}, \mathbf{x}_{s1}) & q^*(\mathbf{x}_{N_1}, \mathbf{x}_{s2}) & \cdots & q^*(\mathbf{x}_{N_1}, \mathbf{x}_{sN}) \end{bmatrix} \begin{bmatrix} \alpha_1 \\ \alpha_2 \\ \vdots \\ \alpha_N \end{bmatrix} = \begin{bmatrix} f(\mathbf{x}_1) \\ \vdots \\ f(\mathbf{x}_{N_1}) \\ \hline g(\mathbf{x}_1) \\ \vdots \\ g(\mathbf{x}_{N_2}) \end{bmatrix} \tag{1.46}$$

or

$$\mathbf{H}\boldsymbol{\alpha} = b \tag{1.47}$$

from which the approximating coefficients can be solved and then the quantities at any point in the domain can be evaluated by Eqs. (1.37) and (1.41) for further plotting or analysis.

The unique existence of approximate solutions for Eqs. (1.31) and (1.33) of the form (1.37) has been established theoretically. However, the locations of source points $\{\mathbf{x}_{sk}\}_{k=1}^{N}$ outside the computing domain cannot be explicitly determined. In practical computation, we can usually employ the following relation to generate the source point:

$$\mathbf{x}_s = \mathbf{x}_b + (\mathbf{x}_b - \mathbf{x}_c)\gamma \tag{1.48}$$

where $\mathbf{x}_s$, $\mathbf{x}_b$, and $\mathbf{x}_c$ represent the positions of source point, boundary point, and center point, respectively. γ denotes the nondimensional parameter used to control the distance between the source point and the real boundary.

Obviously, Eq. (1.48) can be equally rewritten as

$$\mathbf{x}_s - \mathbf{x}_c = \mathbf{x}_b - \mathbf{x}_c + (\mathbf{x}_b - \mathbf{x}_c)\gamma = (\mathbf{x}_b - \mathbf{x}_c)(1 + \gamma) \tag{1.49}$$

or

$$\mathbf{x}_s = \mathbf{x}_c + (\mathbf{x}_b - \mathbf{x}_c)\lambda \tag{1.50}$$

where $\lambda = 1 + \gamma > 0$ is usually called the similarity ratio between the virtual boundary source points located and the real physical boundary.

Eq. (1.50) means that the coordinates of the source point generated by the relation (1.48) are scaled magnification or shrinkage of the coordinates of the boundary collocation point. Theoretically, for the case of the interior domain (see Fig. 1.7), the scaling factor $\gamma > 0$, whereas for the case of the exterior domain (see Fig. 1.8), the parameter γ can be chosen in the range $(-1, 0)$. For both the interior and external domain problems, it is found that the closer the parameter γ is to 0, the closer the virtual boundary connecting the source points is to the physical boundary. However, from the computational point of view, the numerical accuracy may decrease as the distance between the virtual

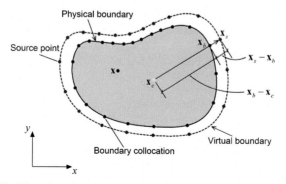

FIGURE 1.7   Illustration of the locations of source points for interior domain problems.

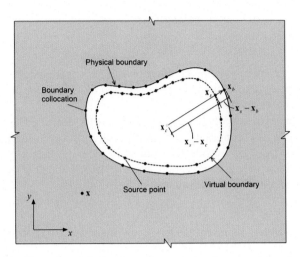

FIGURE 1.8   Illustration of the locations of source points for external domain problems.

and physical boundaries becomes too small, in which case the parameter γ is close to zero, and the problem becomes singular due to the singularity of the fundamental solutions. In contrast, as the distance of source points from the physical boundary increases, the conditioning number of the collocation matrix of the linear solving system of equations increases, and the round-off errors in floating-point arithmetic may become a concern. Typically, if the parameter γ becomes very large, the entries in the collocation matrix will be close to zero. Therefore, to achieve a good balance for the conditioning number and the numerical errors generated by the MFS in practical computation, the parameter γ is generally selected to be in the range of $1.0-10.0$ for interior domain problems and $-0.3 \sim -0.8$ for external domain problems [34,42,45,46]. In practice, we usually employ $\gamma = 3$ for the interior case and $\gamma = -0.5$ for the exterior case.

1.4.3 Program structure and source code

Based on the MFS formulation described before, MATLAB code can be programmed for the numerical analysis of two-dimensional Laplace problems. The main solution procedures of the program include the following:

- Read input data and allocate proper array sizes.
- **For** each boundary collocation **do**:
  - Compute coefficients;
  - Assemble to form coefficient matrix.
- Solve the resulting matrix equation for the approximating coefficients.
- Compute results at test points.
- Output the potential and its derivative solutions.

1.4.3.1 Input data

The major input parameters needed for the implementation of MFS are these:

ND: number of dimensions of problem
NF: number of degrees of freedom (DOFs) per point
NR: number of boundary collocations on the real boundary
NV: number of virtual source points
NT: number of test points in the domain
RC: NR by ND matrix storing coordinates of each boundary collocation
RN: NR by ND matrix storing normals at each boundary collocation
BT: NR by NF matrix storing type of boundary condition at each boundary collocation
BV: NR by NF matrix storing given value of boundary condition at each boundary collocation
VC: NV by ND matrix storing coordinates of virtual source points
TC: NT by ND matrix storing coordinates of test points

Most of these parameters input from data files are associated with the geometric information about the solution domain and its given boundary conditions of a problem defined by a user. The boundary collocations can be generated either manually or using an automatic boundary discretization program (i.e., the preprocessor in the BEM), and they can then be used to generate source points outside the domain using Eq. (1.48). Additionally, the information for the given boundary condition at each boundary collocation includes the type of boundary condition (BT) and the given constraint value (BV). For the present problem, the degree of freedom of each boundary collocation is NF = 1, so the information for any constrained boundary collocation i (i = 1, 2, ..., NR) is as follows:

BT[i,1] = 0: specified potential
BT[i,1] = 1: specified normal flux
BV[i,1] = specified value

1.4.3.2 Computation of coefficient matrix

The resulting coefficient matrix **H** is expressed in Eq. (1.46). The size of the **H** matrix is determined by the number of boundary collocations and source points prepared in the input part, and each element in it is computed from the fundamental solution u^* or q^* given in Eqs. (1.34) and (1.43) with the coordinates of the specific boundary collocation and source point.

1.4.3.3 Solving the resulting system of linear equations

Once the system of linear equations is established, it can be solved for the approximating coefficients α_i, which are the primary unknowns in the MFS.

In the MATLAB program, such systems can be solved by the inbuilt function, that is, the matrix left division,

$$\boldsymbol{\alpha} = H \backslash \mathbf{b}$$

Finally, the physical quantities at any point in the domain can be evaluated with the solved coefficients α_i.

1.4.3.4 Source code

Source code is written in M-function in MATLAB, and the main function MFS_2DLaplace.m calls another M-function (subroutine) FDS2DLP.m that is coded to evaluate the fundamental solution u^* and its derivatives $\partial u^*/\partial x$ and $\partial u^*/\partial y$ when the coordinates of source point and field point are input. The flowchart of the program is given in Fig. 1.9.

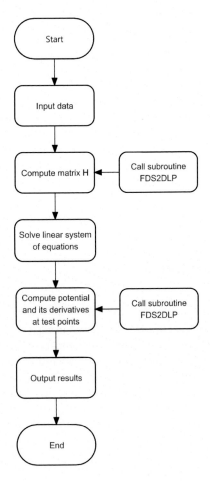

FIGURE 1.9 Flowchart of the program of MFS_2DLaplace.m for solving two-dimensional Laplace problems.

```
% ====================================================
% Main program to solve 2D Laplace problem using the MFS

% ====================================================
function MFS_2DLaplace
%

% **** Laplace equation:

%               Uxx + Uyy = 0
% **** Boundary conditions
%   ** Dirichlet: U=U0
%   ** Neumann:    Q=Q0
% where
%   U is the sought potential function
%   Q is the normal derivative of potential U, that is Q=dU/dn
%
% **** Variable statements
%   ND: Number of dimensions of problem
%   NF: Number of DOFs of problem at every point
%   NR: Number of boundary collocations on the real boundary
%
%   NV: Number of virtual source points
%   NT: Number of test points in the domain
%
%   RC: NR by ND matrix of coordinates of boundary collocations
%   RN: NR by ND matrix of normals at boundary collocations
%   BT: NR by NF matrix of types of boundary conditions at boundary
%       collocations
%   BV: NR by NF matrix of values of boundary conditions
%   VC: NV by ND matrix of coordinates of virtual source points
%   TC: NT by ND matrix of coordinates of test points
%
%---------------------------------------------------------------
% ** Open data file for input
fp=fopen('Input.txt','rt');
dummy=char(zeros(1,100));
% Test description
dummy=fgets(fp);
% Basic parameters
dummy=fgets(fp);
TMP=str2num(fgets(fp));
[ND,NF,NR,NV,NT]=deal(TMP(1),TMP(2),TMP(3),TMP(4),TMP(5));
% Initialization
RC=zeros(NR,ND);
RN=zeros(NR,ND);
```

```
    TB=zeros(NR,NF);
    VB=zeros(NR,NF);
    VC=zeros(NV,ND);
    TC=zeros(NT,ND);
    % Boundary collocation coordinates and normals
    dummy=fgets(fp);
    dummy=fgets(fp);
    for i=1:NR
       TMP=str2num(fgets(fp));
       [NUM,RC(i,1:ND),RN(i,1:ND)]=
deal(TMP(1),TMP(2:1+ND),TMP(2+ND:1+2*ND));
    end
    % Type and value of boundary conditions
    dummy=fgets(fp);
    dummy=fgets(fp);
    for i=1:NR
       TMP=str2num(fgets(fp));
       [NUM,BT(i,1:NF),BV(i,1:NF)]=
deal(TMP(1),TMP(2:1+NF),TMP(2+NF:1+2*NF));
    end
    % Source points
    dummy=fgets(fp);
    dummy=fgets(fp);
    for i=1:NV
      TMP=str2num(fgets(fp));
       [NUM,VC(i,1:ND)]=deal(TMP(1),TMP(2:1+ND));
    end
    % Test points
    dummy=fgets(fp);
    dummy=fgets(fp);
    for i=1:NT
      TMP=str2num(fgets(fp));
       [NUM,TC(i,1:ND)]=deal(TMP(1),TMP(2:1+ND));
    end
    fclose(fp);

    % ** Form the coefficient matrix H
    HH=zeros(NR,NV);
    % Satisfy boundary conditions at NR boundary collocations
    for i=1:NR

      x=RC(i,1);
      y=RC(i,2);
      nx=RN(i,1);
      ny=RN(i,2);
      for j=1:NV
```

```
        vx=VC(j,1);
        vy=VC(j,2);
        [h,hx,hy]=FDS2DLP(x,y,vx,vy);
        q=(hx*nx+hy*ny);
        if BT(i,1)==0 % Specified potential
            HH(i,j)=h;
        else             % Specified normal derivative
            HH(i,j)=q;
        end
    end
end
%-------------------------------------------------------------------
% ** Solve the linear system of equations
BV=HH\BV; % Solve Ax=b using matrix left division
%-------------------------------------------------------------------
% **Evaluate quantities at test points
UC=zeros(NT,3);
for i=1:NT
    x=TC(i,1);
    y=TC(i,2);
    for j=1:NV
        vx=VC(j,1);
        vy=VC(j,2);
        [h,hx,hy]=FDS2DLP(x,y,vx,vy);
        UC(i,1)=UC(i,1)+BV(j,1)*h;
        UC(i,2)=UC(i,2)+BV(j,1)*hx;
        UC(i,3)=UC(i,3)+BV(j,1)*hy;
    end
end
%-------------------------------------------------------------------
%**Output results in Cartesian coordinate system at test points
fp=fopen('Results.txt','wt');
fprintf(fp,'No.    x,    y,    U,    Q1,    Q2 \n');
for i=1:NT
    x=TC(i,1);
    y=TC(i,2);
    u =UC(i,1);
    q1=UC(i,2);
    q2=UC(i,3);
    fprintf(fp,'%3d, %5.3e, %5.3e, %9.6e, %9.6e, %9.6e\n',i,x,y,u,
q1,q2);
end
fclose(fp);
```

```
close all;
%------------End main program-----------------
% ===============================================
% Subroutine FDS2DLP: Compute the 2D fundamental solution at the field
  point
% (x,y) with the given source point (vx,vy)
% ===============================================
function [h,hx,hy]=FDS2DLP(x,y,vx,vy)
%
rx=x-vx;
ry=y-vy;
r=sqrt(rx^2+ry^2);
rx=rx/r; % dr/dx
ry=ry/r; % dr/dy
h=log(1.0/r)/(2*pi);          % u*
hx=-(1.0/2/pi/r)*rx;          % du*/dx
hy=-(1.0/2/pi/r)*ry;          % du*/dy
```

1.4.4 Numerical experiments

To demonstrate the performance and applicability of the MFS for two-dimensional Laplace problems, three examples, a simple circular disk, a simply connected domain with a complex boundary, and a hollow circle (annulus), are considered. Additionally, the following averaged relative error norm on physical quantity ζ is defined as an error indicator to assess the accuracy of numerical results:

$$\text{Arerr}(\zeta) = \frac{\sqrt{\sum_{i=1}^{L} \left(\zeta_i - \tilde{\zeta}_i\right)^2}}{\sqrt{\sum_{i=1}^{L} \zeta_i^2}} \tag{1.51}$$

where L is the total number of test points in the entire computing domain, ζ_i and $\tilde{\zeta}_i$ are the exact and numerical results of variable ζ at the test point $\mathbf{x}_i$ ($i = 1, 2, ..., L$) of interest, respectively.

1.4.4.1 Circular disk

We first consider the case where the computing domain is a circular disk with unit radius, as illustrated in Fig. 1.10. It is assumed that the nonlinear function $u = e^x \cos y$ is adopted to produce the Dirichlet boundary condition along the circular edge. It can be easily checked that $u = e^x \cos y$ exactly satisfies Eq. (1.31). Figs. 1.11 and 1.12 show the distributions of the exact solution and its derivative to the spatial variable y in the circular disk. Here, $\partial u/\partial x$ is same as u and thus is not plotted.

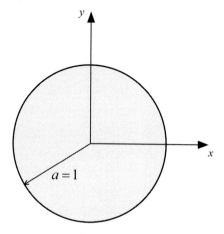

FIGURE 1.10   Schematic illustration of *circular disk* with unit radius.

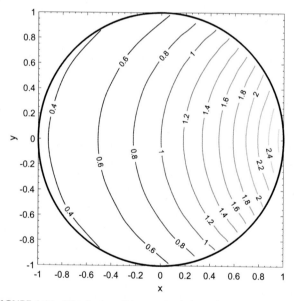

FIGURE 1.11   Distribution of the exact solution u in the *circular disk*.

For such circular domains, the source points and boundary collocations can be chosen regularly, as depicted in Fig. 1.13. Besides, to investigate the numerical accuracy of the MFS, the results at 85 test points $\{\mathbf{x}_k\}_{k=1}^{M}$ displayed in Fig. 1.14 are evaluated.

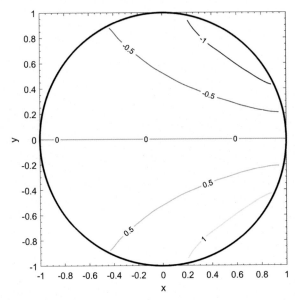

FIGURE 1.12 Distribution of the exact solution $\partial u/\partial y$ in the *circular disk*.

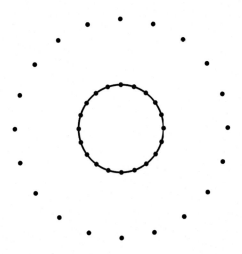

FIGURE 1.13 Schematic illustration of source points and boundary collocations for the *circular disk* domain.

First, the convergence of the MFS is investigated. Figs. 1.15, 1.16, and 1.17 indicate the error results of u and its derivatives at the selected 85 test points to the number of collocations N on the boundary. It is found that the both the potential field u and its derivatives give a high order of convergence for the MFS scheme. On the other hand, the distributions of the numerical results at

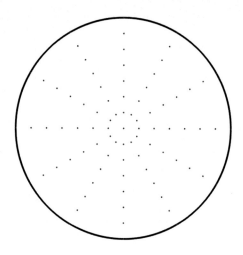

FIGURE 1.14 Schematic illustration of test points in the *circular disk* domain.

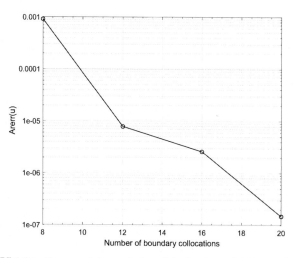

FIGURE 1.15 Convergent demonstration of the function *u* for the *circular disk*.

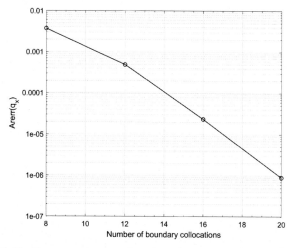

FIGURE 1.16 Convergent demonstration of the derivative ∂*u*/∂*x* for the *circular disk*.

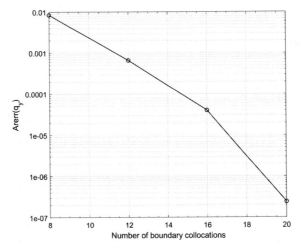

FIGURE 1.17   Convergent demonstration of the derivative $\partial u/\partial y$ for the *circular disk*.

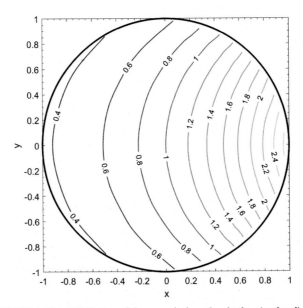

FIGURE 1.18   Distribution of the numerical result u in the *circular disk*.

the test points in the computing domain are plotted for comparison with the exact distributions. From the results depicted in Figs. 1.18–1.20, it can be observed that the MFS accurately captures the complex variations of the potential field and its derivatives in the circular disk, and the same distributions of u and $\partial u/\partial x$ are produced by the MFS, also demonstrating the effectiveness of the MFS.

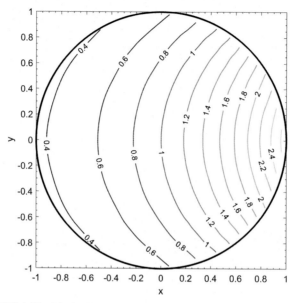

FIGURE 1.19   Distribution of the numerical result $\partial u / \partial x$ in the *circular disk*.

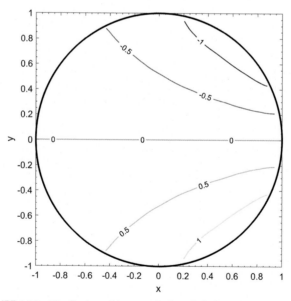

FIGURE 1.20   Distribution of the numerical result $\partial u / \partial y$ in the *circular disk*.

1.4.4.2 Interior region surrounded by a complex curve

We next consider the case (see Fig. 1.21) in which the boundary is depicted by the following function:

$$\partial \Omega = \left\{ (r \cos \theta, \ r \sin \theta) : \ r = e^{\sin \theta} \sin^2(2\theta) + e^{\cos \theta} \cos^2(2\theta), \ 0 \le \theta \le 2\pi \right\}$$

$$(1.52)$$

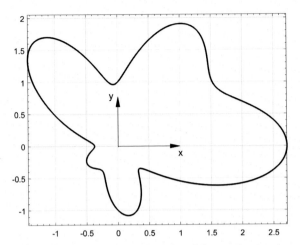

FIGURE 1.21 Geometric configuration of the complex domain.

In the computation, 30 collocation points are arranged along the boundary $\partial\Omega$ by setting θ to change uniformly at the intervals $(0, 2\pi)$. Then, the related singular source points can be generated by Eq. (1.48) with $\gamma = 3$, as displayed in Fig. 1.22. As well, 134 test points are selected to evaluate the related potential u and its derivatives at those points, as shown in Fig. 1.23.

Figs. 1.24 and 1.25 depict the analytical distributions of the exact solutions for u and its derivative $\partial u/\partial y$, respectively. It is clearly seen that the potential u and its derivative have complex distributions in the domain with the complex boundary shape. With the developed MFS program, numerical results at 134 test points are evaluated and plotted in Figs. 1.26–1.28, from which good

FIGURE 1.22 Schematic illustration of source points and collocation points for the complex domain.

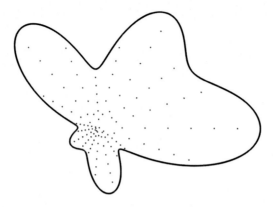

FIGURE 1.23 Configurations of test points in the complex domain.

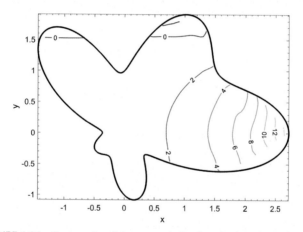

FIGURE 1.24 Contour plot of the exact solution for u in the complex domain.

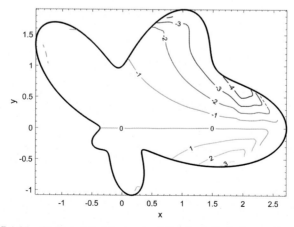

FIGURE 1.25 Contour plot of the exact solution for $\partial u/\partial y$ in the complex domain.

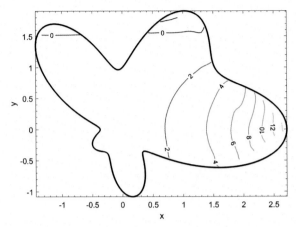

FIGURE 1.26   Contour plot of numerical result for u in the complex domain.

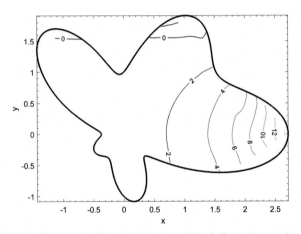

FIGURE 1.27   Contour plot of the numerical result for $\partial u/\partial x$ in the complex domain.

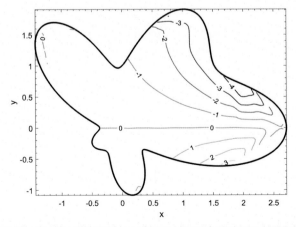

FIGURE 1.28   Contour plot of the numerical result for $\partial u/\partial y$ in the complex domain.

agreement is found with the analytical distributions in Figs. 1.24 and 1.25. Moreover, the correspondence of the numerical results of u and $\partial u/\partial x$ in Figs. 1.26 and 1.27 also demonstrates the correctness of the numerical results and the applicability of the MFS.

1.4.4.3 Biased hollow circle

In the third example, a biased hollow circle is considered, as shown in Fig. 1.29. As before, the exact solution $u = e^x \cos(y)$ is used to apply the Dirichlet boundary condition along the outer and inner circular edges. Sixteen collocation points are uniformly arranged along the outer boundary and 12 collocation points along the inner boundary. Then, the related singular source points can be generated by Eq. (1.48) with $\gamma = 3$ for the exterior circular boundary and $\gamma = -0.5$ for the interior circular boundary, as displayed in Fig. 1.30. As well, 189 test points are chosen for further result treatment, as shown in Fig. 1.31.

For comparison, the exact solutions for u and the derivative $\partial u/\partial y$ are plotted in Figs. 1.32 and 1.33. With the developed MFS program, the numerical results at the test points are obtained and then plotted in Figs. 1.34–1.36. Good agreement between the numerical and exact solutions is observed for this multiply connected domain problem. In addition, Figs. 1.34 and 1.35 related to the numerical results of the variables u and $\partial u/\partial x$ respectively show almost the same distribution. This result also verifies the effectiveness of the MFS.

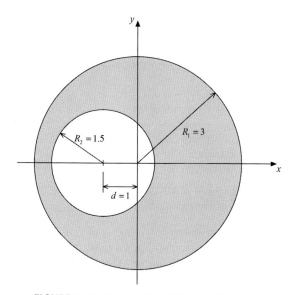

FIGURE 1.29 Configuration of biased *hollow circle*.

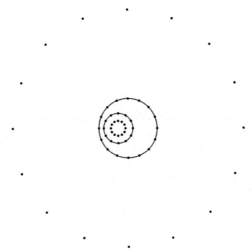

FIGURE 1.30 Locations of source and collocation points for the biased *hollow circle*.

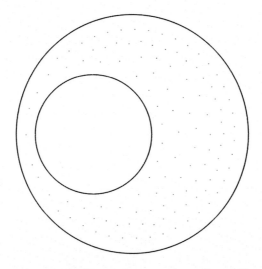

FIGURE 1.31 Configuration of test points for the biased *hollow circle*.

1.5 Some limitations for implementing the method of fundamental solutions

From the solution procedure of the MFS described here, we can conclude that the MFS formulation is very simple, and its implementation is direct. More importantly, the MFS has very high numerical accuracy and rapid convergence [40–42]. However, it should be noted that the MFS has some disadvantages that may limit its application [47].

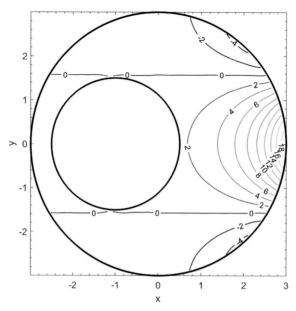

FIGURE 1.32   Contour plot of the exact solution u in the biased *hollow circle*.

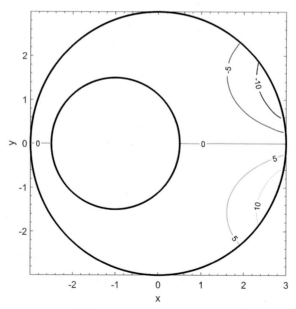

FIGURE 1.33   Contour plot of the exact solution $\partial u/\partial y$ in the biased *hollow circle*.

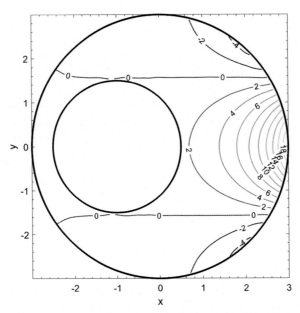

FIGURE 1.34 Contour plot of the numerical solution u in the biased *hollow circle*.

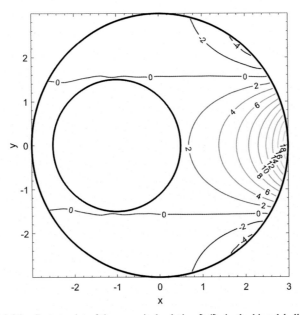

FIGURE 1.35 Contour plot of the numerical solution $\partial u/\partial x$ in the biased *hollow circle*.

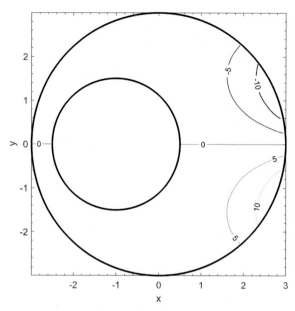

FIGURE 1.36 Contour plot of the numerical solution $\partial u/\partial y$ in the biased *hollow circle*.

1.5.1 Dependence of fundamental solutions

In the MFS, the fundamental solutions of a problem are used to construct the approximation of the field variable of interest so that the governing equations of the problem can be analytically satisfied. This means that implementation of the MFS depends heavily on the fundamental solutions of the problem. If the fundamental solutions of the problem are predefined, the MFS may be a good choice for finding numerical solutions. For some problems, however, that may be linear or nonlinear, it is difficult to derive the corresponding fundamental solutions. For such cases, the MFS cannot be applied directly. To overcome this obstacle, the analog equation method developed by Katsikadelis in the BEM [19] can be employed to convert a problem without explicit fundamental solutions available to an equivalent linear inhomogeneous Poisson's problem, whose fundamental solutions are typically known. During the past decade, this strategy has been successfully applied for solving problems with variable coefficients [45,48,49], which usually occur in inhomogeneous materials such as functionally graded materials, and in nonlinear problems [46,50].

1.5.2 Location of source points

One of the main difficulties in the application of the MFS is determination of the position of source points to obtain an optimal solution, and this issue is still open for researchers [51].

Usually, the locations of source points are determined by considering either a static scheme, in which the locations of sources are preassigned [34,45,46], as shown in Section 1.4, or a dynamic scheme, in which both the locations of sources and the unknown coefficients $\{\alpha_j\}_{j=1 \to N}$ are simultaneously determined during the solution procedure [34,52]. The dynamic scheme causes a nonlinear system to be solved and entails unavoidable difficulty in solving. The static scheme is more easily implemented.

The most frequently used static schemes involve assigning the source points along either a circular virtual boundary or a geometrically similar virtual boundary, as indicated in Fig. 1.37. In this book, the static scheme with a similar virtual boundary is preferred [45,46,53,54], and the source point can be generated by the simple relation (1.48).

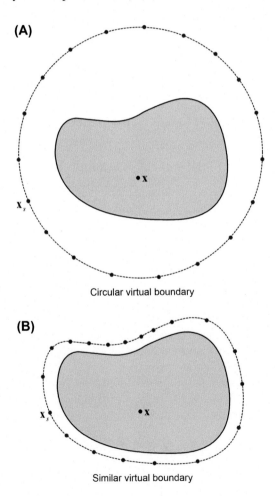

(A)

Circular virtual boundary

(B)

Similar virtual boundary

FIGURE 1.37   Circular and similar virtual boundaries.

1.5.3 Ill-conditioning treatments

It is well known that the ill-conditioning of the MFS matrix $\mathbf{A}^T\mathbf{A}$ increases as the distance between the source points $(\mathbf{x}_{sj})_{j=1 \to N}$ and the physical boundary $\partial\Omega$ increases, or as the number of boundary collocations increases [14,55]. Thus, in such cases, the system of Eq. (1.29) needs to be specially dealt with. To illustrate this procedure conveniently, the following general linear system of equations is considered.

$$\mathbf{A}\boldsymbol{\alpha} = \mathbf{f} \tag{1.53}$$

where $\mathbf{A}$ is a square matrix, $\boldsymbol{\alpha}$ and $\mathbf{f}$ are vectors, respectively.

1.5.3.1 Tikhonov regularization method

The first treatment is the Tikhonov regularization method [56,57], which gives

$$\boldsymbol{\alpha}_\lambda = (\mathbf{A} + \lambda\mathbf{I})^{-1}\mathbf{f} \tag{1.54}$$

where $\mathbf{I}$ is the identity matrix, and $\lambda > 0$ is a regularization parameter to be prescribed according to some criterion such as the L-curve criterion.

Eq. (1.54) imposes a continuity constraint onto the solution $\boldsymbol{\alpha}$ and is known as Tikhonov's regularization of order zero, which actually can be derived from minimization of the functional

$$\Pi(\boldsymbol{\alpha}) = \|\mathbf{A}\boldsymbol{\alpha} - \mathbf{f}\|^2 + \lambda\|\boldsymbol{\alpha}\| \tag{1.55}$$

1.5.3.2 Singular value decomposition

Alternatively, instead of the Tikhonov regularized solution, the truncated singular value decomposition method [58,59] can be employed to treat the ill-conditioning of the coefficient matrix.

According the theory of singular value decomposition, the matrix $\mathbf{A}$ can be written as the product a column-orthogonal matrix $\mathbf{U}$, a diagonal matrix $\mathbf{W}$ with nonnegative elements (the singular values), and the transpose of an orthogonal matrix $\mathbf{V}$, that is

$$\mathbf{A} = \mathbf{U}\mathbf{W}\mathbf{V}^T = w_1\mathbf{u}_1\mathbf{v}_1^T + w_2\mathbf{u}_2\mathbf{v}_2^T + \ldots w_r\mathbf{u}_r\mathbf{v}_r^T \tag{1.56}$$

where

$$\begin{aligned}
\mathbf{U} &= \begin{bmatrix} \mathbf{u}_1 & \mathbf{u}_2 & \cdots & \mathbf{u}_r \end{bmatrix} \\
\mathbf{V} &= \begin{bmatrix} \mathbf{v}_1 & \mathbf{v}_2 & \cdots & \mathbf{v}_r \end{bmatrix}
\end{aligned} \tag{1.57}$$

and

$$\mathbf{W} = \mathrm{diag}(w_1, w_2, \cdots w_r) = \begin{bmatrix} w_1 & 0 & \cdots & 0 \\ 0 & w_2 & \cdots & 0 \\ \vdots & \vdots & \ddots & \vdots \\ 0 & 0 & \cdots & w_r \end{bmatrix} \tag{1.58}$$

with the singular value $w_1 \geq w_2 \geq \ldots \geq w_r > 0$.

As a result, the solution of (1.53) can be computed by

$$\boldsymbol{\alpha} = \mathbf{V} \cdot [\text{diag}(1/w_j)] \cdot (\mathbf{U}^T\mathbf{f}) \tag{1.59}$$

Formally, the condition number of a matrix is defined as the ratio of the largest singular value to the smallest singular value, that is

$$\text{cond}(\mathbf{A}) = \frac{w_1}{w_r} \tag{1.60}$$

A matrix is singular if its condition number is infinite, and it is ill-conditioned if its condition number is too large or if its reciprocal approaches the machine's floating-point precision.

In the case of an ill-conditioned matrix, the direct solution methods of lower–upper (LU) decomposition or Gaussian elimination actually yield a formal solution to the set of equations; that is, a zero pivot may not be encountered, but the solution vector may have extremely large components whose algebraic cancellation, when multiplied by the matrix $\mathbf{A}$, may give a very poor approximation of the right-hand vector $\mathbf{f}$. In such a case, the solution vector obtained by zeroing the smaller singular values may be better (in the sense of the residual $|\mathbf{A}\boldsymbol{\alpha} - \mathbf{f}|$ being smaller) than both the direct-method solution and the regular singular value decomposition solution where the smaller singular values are left as nonzero [60]. For instance, if the singular values w_k ($k = k_1 \rightarrow r$) are very small, we can set $w_k = 0$. Then, the diagonal matrix $\mathbf{W}$ can be rewritten as

$$\mathbf{W} = \begin{bmatrix} w_1 & 0 & \cdots & 0 & 0 & \cdots & 0 \\ 0 & w_2 & \cdots & 0 & 0 & \cdots & 0 \\ \vdots & \vdots & \ddots & \vdots & \vdots & \ddots & \vdots \\ 0 & 0 & \cdots & w_{k1-1} & & \cdots & 0 \\ 0 & 0 & \cdots & 0 & 0 & \cdots & 0 \\ \vdots & \vdots & \ddots & \vdots & \vdots & \ddots & \vdots \\ 0 & 0 & \cdots & 0 & 0 & \cdots & 0 \end{bmatrix} \tag{1.61}$$

As a result, the solution of (1.53) can be computed by

$$\boldsymbol{\alpha} = \mathbf{V} \cdot \begin{bmatrix} \dfrac{1}{w_1} & 0 & \cdots & 0 & 0 & \cdots & 0 \\ 0 & \dfrac{1}{w_2} & \cdots & 0 & 0 & \cdots & 0 \\ \vdots & \vdots & \ddots & \vdots & \vdots & \ddots & \vdots \\ 0 & 0 & \cdots & \dfrac{1}{w_{k1-1}} & \vdots & \cdots & 0 \\ 0 & 0 & \cdots & 0 & 0 & \cdots & 0 \\ \vdots & \vdots & \ddots & \vdots & \vdots & \ddots & \vdots \\ 0 & 0 & \cdots & 0 & 0 & \cdots & 0 \end{bmatrix} \cdot (\mathbf{U}^T\mathbf{f}) \tag{1.62}$$

In the following sections of this book, the truncated singular value decomposition method is preferably employed to deal with a high condition number of the coefficient matrix in the final solving system of equations.

1.5.4 Inhomogeneous problems

Inhomogeneous problems in general refer to those with a nonzero right-hand term in the governing partial differential equation of the problem. The nonzero right-hand term may be led to by a source term in the domain, such as the internal thermal generation per volume, body forces, inertial forces, etc. Also, for time-dependent problems, time discretization can also bring inhomogeneous terms into the governing equations. For such problems, the MFS cannot be applied directly, because the approximate solutions in the MFS satisfy only homogeneous governing equations. Fortunately, the radial basis function (RBF) interpolation can be employed coupled with the MFS to overcome this disadvantage. In the 1990s, for example, trials were made to deal with inhomogeneous Poisson's equations and time-dependent problems, with good effect [34,61,62].

Nevertheless, it is necessary to point out that coupling of the MFS and the RBF interpolation requires interior collocations in the domain. Such treatment may lose some of the advantages of the standard MFS.

1.5.5 Multiple domain problems

In engineering, some problems include multiple material subdomains [54]. In such cases, the standard MFS requires separate boundary collocation discretization for each subdomain. Then the continuity condition of variables along the common interface should be satisfied to connect the separate MFS formulation of each subdomain.

To illustrate this issue in detail, we consider a two-dimensional static potential problem in a domain with two different material definitions, as shown in Fig. 1.38. It is assumed that each subdomain is isotropic and homogeneous. The potential field is generally governed by the standard potential equation

$$\mathbf{L}u = 0 \tag{1.63}$$

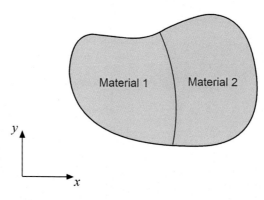

FIGURE 1.38 Problem with two different material definitions.

For simplicity, it is assumed that the outer boundary is isothermal, that is, $u|_\Gamma = \bar{u}_0$. Moreover, the free-space fundamental solution for such potential equations is assumed to be $G^*(\mathbf{x}, \mathbf{y})$, in which $\mathbf{x}$ is the position of the field point and $\mathbf{y}$ is the position of the source point.

Because the MFS uses the boundary values only for the interior domain solution, the individual subdomains are solved easily with known boundary conditions. In the present case, the MFS formulation for subdomain 1 with material 1 can be established by boundary collocation discretization:

$$u^{(1)}\left(\mathbf{x}_i^{(1)}, \mathbf{y}_j^{(1)}\right) = \sum_{j=1}^{N^{(1)}} \alpha_j^{(1)} G^*\left(\mathbf{x}_i^{(1)}, \mathbf{y}_j^{(1)}\right) \tag{1.64}$$

$$u_n^{(1)}\left(\mathbf{x}_i^{(1)}, \mathbf{y}_j^{(1)}\right) = \frac{\partial u^{(1)}\left(\mathbf{x}_i^{(1)}, \mathbf{y}_j^{(1)}\right)}{\partial n} = \sum_{j=1}^{N^{(1)}} \alpha_j^{(1)} \frac{\partial G^*\left(\mathbf{x}_i^{(1)}, \mathbf{y}_j^{(1)}\right)}{\partial n} \tag{1.65}$$

while for subdomain 2 with material 2, the approximate solution can be obtained as shown:

$$u^{(2)}\left(\mathbf{x}_i^{(2)}, \mathbf{y}_j^{(2)}\right) = \sum_{j=1}^{N^{(2)}} \alpha_j^{(2)} G^*\left(\mathbf{x}_i^{(2)}, \mathbf{y}_j^{(2)}\right) \tag{1.66}$$

$$u_n^{(2)}\left(\mathbf{x}_i^{(2)}, \mathbf{y}_j^{(2)}\right) = \frac{\partial u^{(2)}\left(\mathbf{x}_i^{(2)}, \mathbf{y}_j^{(2)}\right)}{\partial n} = \sum_{j=1}^{N^{(2)}} \alpha_j^{(2)} \frac{\partial G^*\left(\mathbf{x}_i^{(2)}, \mathbf{y}_j^{(2)}\right)}{\partial n} \tag{1.67}$$

where $G^*(\mathbf{x}_i, \mathbf{y}_j)$ is the fundamental solution of the i-th boundary node and the j-th source point, α_j are the undetermined coefficients, n is the unit normal at the boundary node $\mathbf{x}_i$, and N is the number of source points. The superscripts (1) and (2) represent subdomains 1 and 2, respectively.

Obviously, each separation domain includes two individual boundaries: that is, $\Gamma_1^{(1)}$ and $\Gamma_2^{(1)}$ belong to subdomain 1, and $\Gamma_1^{(2)}$ $\Gamma_2^{(2)}$ belong to subdomain 2, as shown in Fig. 1.39. From Eqs. (1.64)−(1.67), the approximations

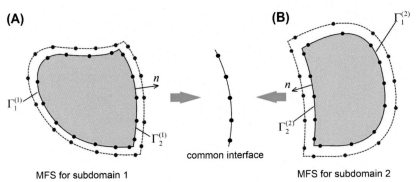

FIGURE 1.39 MFS discretization for problem with two different subdomains.

using the MFS are expressed in matrix form for the following individual subdomains:

$$
\begin{bmatrix} \mathbf{A}^{(1)} \\ \mathbf{B}^{(1)} \\ \mathbf{C}^{(1)} \end{bmatrix} \boldsymbol{\alpha}^{(1)} = \left\{ \begin{array}{c} \bar{\mathbf{u}}_0 \big|_{\Gamma_1^{(1)}} \\ \mathbf{u} \big|_{\Gamma_2^{(1)}} \\ \mathbf{u}_n \big|_{\Gamma_2^{(1)}} \end{array} \right\} \tag{1.68}
$$

and

$$
\begin{bmatrix} \mathbf{A}^{(2)} \\ \mathbf{B}^{(2)} \\ \mathbf{C}^{(2)} \end{bmatrix} \boldsymbol{\alpha}^{(2)} = \left\{ \begin{array}{c} \bar{\mathbf{u}}_0 \big|_{\Gamma_1^{(2)}} \\ \mathbf{u} \big|_{\Gamma_2^{(2)}} \\ \mathbf{u}_n \big|_{\Gamma_2^{(2)}} \end{array} \right\} \tag{1.69}
$$

Moreover, it is required that the following connection conditions at the common nodes on the interface Γ_2 of the subdomains be satisfied:

$$
\begin{aligned}
\mathbf{u}\big|_{\Gamma_2^{(1)}} &= \mathbf{u}\big|_{\Gamma_2^{(2)}} \\
\mathbf{u}_n\big|_{\Gamma_2^{(1)}} &= \mathbf{u}_n\big|_{\Gamma_2^{(2)}}
\end{aligned} \tag{1.70}
$$

By combining from Eq. (1.68) to Eq. (1.70), the final solving system of global linear equations for the entire computational domain can be obtained as follows:

$$
\begin{bmatrix} \mathbf{A}^{(1)} & 0 \\ 0 & \mathbf{A}^{(2)} \\ \mathbf{B}^{(1)} & -\mathbf{B}^{(2)} \\ \mathbf{C}^{(1)} & -\mathbf{C}^{(2)} \end{bmatrix} \begin{bmatrix} \boldsymbol{\alpha}^{(1)} \\ \boldsymbol{\alpha}^{(2)} \end{bmatrix} = \begin{bmatrix} \bar{\mathbf{u}}_0\big|_{\Gamma_1^{(1)}} \\ \bar{\mathbf{u}}_0\big|_{\Gamma_1^{(2)}} \\ 0 \\ 0 \end{bmatrix} \tag{1.71}
$$

As a result, it is found that for multiple domain problems, the MFS solution procedure becomes very complex, to the extent that the MFS may lose its advantage over other numerical methods.

1.6 Extended method of fundamental solutions

To overcome some of disadvantages of the MFS, such as the dependence on fundamental solutions and the invalid treatment for inhomogeneous problems, and broaden its applications in engineering, the standard MFS can be extended by introducing other techniques.

To demonstrate this solution strategy, we consider a general potential problem governed by the following general second-order partial differential equation defined in a two-dimensional domain Ω:

$$
\begin{aligned}
A_1(x,y)\frac{\partial^2 u}{\partial x^2} &+ A_2(x,y)\frac{\partial^2 u}{\partial x \partial y} + A_3(x,y)\frac{\partial^2 u}{\partial y^2} \\
&+ A_4(x,y)\frac{\partial u}{\partial x} + A_5(x,y)\frac{\partial u}{\partial y} + A_6(x,y)u + q(x,y) = 0
\end{aligned} \tag{1.72}
$$

where $u(x, y)$ is the sought scalar field, $A_i(x, y)$ are known coefficients, and $q(x, y)$ is a given interior source term.

To write Eq. (1.72) in a compact form, the differential operator

$$\widetilde{L} = A_1(x, y)\frac{\partial^2}{\partial x^2} + A_2(x, y)\frac{\partial^2}{\partial x \partial y} + A_3(x, y)\frac{\partial^2}{\partial y^2}$$
$$+ A_4(x, y)\frac{\partial}{\partial x} + A_5(x, y)\frac{\partial}{\partial y} + A_6(x, y) \tag{1.73}$$

is introduced, so Eq. (1.72) can be rewritten as

$$\widetilde{L}u(\mathbf{x}) + q(\mathbf{x}) = 0 \tag{1.74}$$

where $\mathbf{x} = (x, y)$ is an arbitrary point in the computing domain.

Also, the following generalized boundary condition is considered:

$$\alpha(\mathbf{x})u(\mathbf{x}) + \beta(\mathbf{x})\frac{\partial u(\mathbf{x})}{\partial n} = \chi_0(\mathbf{x}), \quad \mathbf{x} \in \Gamma \tag{1.75}$$

where $\alpha(\mathbf{x})$, $\beta(\mathbf{x})$, and $\chi_0(\mathbf{x})$ are given functions, Γ is the boundary of the computing domain Ω, $\mathbf{n} = \begin{bmatrix} n_x & n_y \end{bmatrix}^T$ is the unit normal vector to the boundary, and

$$\frac{\partial u}{\partial n} = \begin{bmatrix} n_x & n_y \end{bmatrix} \begin{bmatrix} \dfrac{\partial u}{\partial x} \\ \dfrac{\partial u}{\partial y} \end{bmatrix} = \mathbf{n}^T \nabla u \tag{1.76}$$

To write Eq. (1.75) compactly, the following operator is introduced:

$$L_B = \alpha(\mathbf{x}) + \beta(\mathbf{x})\frac{\partial}{\partial n} \tag{1.77}$$

so Eq. (1.75) is rewritten as

$$L_B u(\mathbf{x}) = \chi_0(\mathbf{x}), \quad \mathbf{x} \in \Gamma \tag{1.78}$$

Obviously, the governing Eq. (1.74) has no explicit fundamental solutions. As well, the inhomogeneous term q hinders direct application of the MFS.

To deal with this, the basic concept of the analog equation method (AEM) detailed in literature [19] for the BEM is employed here, i.e., application of a standard Laplacian operator to the unknown field u, which is usually second-order differentiable in practice, yields

$$\nabla^2 u(\mathbf{x}) = \widetilde{b}(\mathbf{x}) \tag{1.79}$$

where $\widetilde{b}$ is the induced fictitious right-hand term.

Due to the unknown u, $\widetilde{b}$ is also unknown. However, such treatment brings about a certain convenience for further treatment. For example, the linearity of the Laplace operator allows us to divide the solution into two parts: (1) the homogeneous solution and (2) the particular solution,

$$u(\mathbf{x}) = u_h(\mathbf{x}) + u_p(\mathbf{x}) \tag{1.80}$$

where the homogeneous solution $u_h(\mathbf{x})$ is required to satisfy

$$\nabla^2 u_h(\mathbf{x}) = 0 \tag{1.81}$$

while the particular solution $u_p(\mathbf{x})$ is required to satisfy

$$\nabla^2 u_p(\mathbf{x}) = \widetilde{b}(\mathbf{x}) \tag{1.82}$$

For the homogeneous solution of Eq. (1.81), the standard MFS can be directly applied to give the following approximation:

$$u_h(\mathbf{x}) = \sum_{i=1}^{N_s} \varphi_i G^*(\mathbf{x}, \mathbf{y}_i), \quad \mathbf{x} \in \Omega, \mathbf{y} \notin \Omega \tag{1.83}$$

where N_s is the number of boundary collocations, φ_i are unknown approximating coefficients, and $G^*(\mathbf{x}, \mathbf{y}_i) = G_i^*(\mathbf{x})$ are related fundamental solutions of the Laplace equation.

In matrix form, Eq. (1.83) can be rewritten as

$$u_h(\mathbf{x}) = \mathbf{G}(\mathbf{x})\boldsymbol{\varphi}, \quad \mathbf{x} \in \Omega, \mathbf{y} \notin \Omega \tag{1.84}$$

with

$$\mathbf{G}(\mathbf{x}) = \begin{bmatrix} G_1^*(\mathbf{x}) & G_2^*(\mathbf{x}) & \cdots & G_{N_s}^*(\mathbf{x}) \end{bmatrix} \tag{1.85}$$

$$\boldsymbol{\varphi} = \begin{bmatrix} \varphi_1 & \varphi_2 & \cdots & \varphi_{N_s} \end{bmatrix}^{\mathrm{T}} \tag{1.86}$$

Simultaneously, to obtain the particular solution related to the arbitrary right-hand term $\widetilde{b}(\mathbf{x})$, the RBF approximation can be considered. To do this, the following two approximations with similar expression are introduced, respectively, for the arbitrary right-hand term $\widetilde{b}(\mathbf{x})$ and the particular solution term $u_p(\mathbf{x})$:

$$\widetilde{b}(\mathbf{x}) \approx \sum_{j=1}^{N_I} \alpha_j \phi_j(\mathbf{x}) \tag{1.87}$$

$$u_p(\mathbf{x}) = \sum_{j=1}^{N_I} \alpha_j \Phi_j(\mathbf{x}) \tag{1.88}$$

where $\phi_j(\mathbf{x})$ and $\Phi_j(\mathbf{x})$, respectively, represent the RBF and the corresponding particular solution kernel, and N_I is the number of collocations in the computing domain Ω, as indicated in Fig. 1.40.

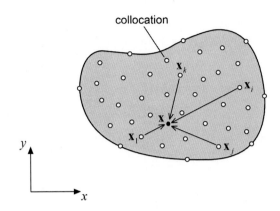

FIGURE 1.40 Collocations in the computing domain for the particular solution.

Eqs. (1.87) and (1.88) can be rewritten in matrix form as follows:

$$\widetilde{b}(\mathbf{x}) = \mathbf{f}(\mathbf{x})\boldsymbol{\alpha} \qquad (1.89)$$

$$u_p(\mathbf{x}) = \boldsymbol{\Phi}(\mathbf{x})\boldsymbol{\alpha} \qquad (1.90)$$

where

$$\boldsymbol{\alpha} = [\,\alpha_1 \quad \alpha_2 \quad \cdots \quad \alpha_{N_I}\,]^{\mathrm{T}} \qquad (1.91)$$

$$\mathbf{f}(\mathbf{x}) = [\,\phi_1(\mathbf{x}) \quad \phi_2(\mathbf{x}) \quad \cdots \quad \phi_{N_I}(\mathbf{x})\,] \qquad (1.92)$$

$$\boldsymbol{\Phi}(\mathbf{x}) = [\,\Phi_1(\mathbf{x}) \quad \Phi_2(\mathbf{x}) \quad \cdots \quad \Phi_{N_I}(\mathbf{x})\,] \qquad (1.93)$$

Substituting Eqs. (1.87) and (1.88) into Eq. (1.82) yields the following connecting equation between $\phi_j(\mathbf{x})$ and $\Phi_j(\mathbf{x})$:

$$\nabla^2 \Phi(r_j) = \phi(r_j) \qquad (1.94)$$

where $\phi(r_j) = \phi_j(\mathbf{x})$, $\Phi(r_j) = \Phi_j(\mathbf{x})$, and $r_j = \|\mathbf{x} - \mathbf{x}_j\|$.

It is clear that once an RBF ϕ is given, integrating Eq. (1.94) in polar coordinates can generate the particular solution kernel Φ. Thus the particular solution expression holds in the form of Eq. (1.88).

Then, the full solution can be written as

$$u(\mathbf{x}) = u_h(\mathbf{x}) + u_p(\mathbf{x}) = \mathbf{G}(\mathbf{x})\boldsymbol{\varphi} + \boldsymbol{\Phi}(\mathbf{x})\boldsymbol{\alpha} = \mathbf{U}(\mathbf{x})\boldsymbol{\beta} \qquad (1.95)$$

where

$$\mathbf{U}(\mathbf{x}) = [\,\mathbf{G}(\mathbf{x}) \quad \boldsymbol{\Phi}(\mathbf{x})\,], \quad \boldsymbol{\beta} = \begin{bmatrix} \boldsymbol{\varphi} \\ \boldsymbol{\alpha} \end{bmatrix} \qquad (1.96)$$

Substituting the full solution (1.95) into the original governing Eq. (1.74) and the boundary condition (1.78) yields

$$\tilde{L}\mathbf{U}(\mathbf{x})\boldsymbol{\beta} + q(\mathbf{x}) = 0, \quad \mathbf{x} \in \Omega \tag{1.97}$$

$$L_B\mathbf{U}(\mathbf{x})\boldsymbol{\beta} = \chi_0(\mathbf{x}), \quad \mathbf{x} \in \Gamma \tag{1.98}$$

Finally, the real satisfaction of the original governing equation at the N_I interpolation points in the domain and the boundary condition at the N_s boundary collocations gives the following linear system of equations in terms of unknown $\boldsymbol{\beta}$:

$$\begin{bmatrix} \tilde{L}\mathbf{U}(\mathbf{x}_1) \\ \vdots \\ \tilde{L}\mathbf{U}(\mathbf{x}_{N_I}) \\ L_B\mathbf{U}(\mathbf{x}_{b1}) \\ \vdots \\ L_B\mathbf{U}(\mathbf{x}_{bN_s}) \end{bmatrix} \boldsymbol{\beta} = \begin{bmatrix} -q(\mathbf{x}_1) \\ \vdots \\ -q(\mathbf{x}_{N_I}) \\ \chi_0(\mathbf{x}_{b1}) \\ \vdots \\ \chi_0(\mathbf{x}_{bN_s}) \end{bmatrix} \tag{1.99}$$

from which the unknown coefficient vector can be determined numerically. Then the full solution at any arbitrary point in the domain can be evaluated by the expression (1.95).

From the preceding solution procedure, it is observed that the introduction of AEM and the RBF can extend the applications of the MFS to more complex problems, and some successive applications of such strategy have been performed for heterogeneous materials, which will cause the appearance of a variable coefficient in the governing equation, and create nonlinear problems [46,48,49,63,64].

1.7 Outline of the book

The aim of this book is to present developments of the MFS, a meshless solution scheme based on fundamental solutions, with which numerical solutions of heat transfer, mechanical deformation, and stresses in some engineering problems can be determined.

Consisting of eight chapters and four appendices, the book covers the fundamentals of continuum mechanics, the basic concepts of MFS, and its methodologies and applications to various engineering problems. In Chapter 1, current meshless methods are reviewed, and the advantages and disadvantages of MFS are outlined. Further, the basic concepts and general formulations of the improved MFS are described. In Chapter 2, the basic mechanical knowledge involved in this book is reviewed, providing a complete source for reference in later chapters. In Chapter 3, the basic concepts of fundamental solutions and RBFs are present. Starting from Chapter 4, some engineering problems, including Euler−Bernoulli beam bending, thin plate bending, plane elasticity or thermoelasticity with or without functionally graded material definitions, plane piezoelectricity, and two-dimensional steady-state or transient heat transfer in heterogeneous materials, are solved in turn by combining the MFS and the RBF interpolation.

The results in this book can improve understanding of the physical and mathematical characteristics of the solution procedures of the improved MFS and may be beneficial to professional engineers, research scientists, and students in engineering.

References

[1] K.J. Bathe, Finite Element Procedures, Prentice-Hall, Inc., New Jersey, 1996.

[2] D.W. Pepper, A.J. Kassab, E.A. Divo, An Introduction to Finite Element, Boundary Element, and Meshless Methods with Applications to Heat Transfer and Fluid Flow, ASME Press, New York, 2014.

[3] Q.H. Qin, C.X. Mao, Coupled torsional-flexural vibration of shaft systems in mechanical engineering—I. Finite element model, Computers and Structures 58 (1996) 835—843.

[4] T.H.H. Pian, C.C. Wu, Hybrid and Incompatible Finite Element Methods, Chapman & Hall/CRC Press, Boca Raton, 2006.

[5] Q.H. Qin, The Trefftz Finite and Boundary Element Method, WIT Press, Southampton, 2000.

[6] Q.H. Qin, H. Wang, MATLAB and C Programming for Trefftz Finite Element Methods, CRC Press, New York, 2009.

[7] Q.H. Qin, Trefftz finite element method and its applications, Applied Mechanics Reviews 58 (2005) 316—337.

[8] Q.H. Qin, Fundamental solution based finite element method, Journal of Applied Mechanical Engineering 2 (2013) e118.

[9] A.H.D. Cheng, D.T. Cheng, Heritage and early history of the boundary element method, Engineering Analysis with Boundary Elements 29 (2005) 268—302.

[10] C.A. Brebbia, J. Dominguez, Boundary Elements: An Introductory Course, Computational Mechanics Publications, Southampton, 1992.

[11] Q.H. Qin, Material properties of piezoelectric composites by BEM and homogenization method, Composite Structures 66 (2004) 295—299.

[12] Q.H. Qin, Nonlinear analysis of Reissner plates on an elastic foundation by the BEM, International Journal of Solids and Structures 30 (1993) 3101—3111.

[13] D.A. Anderson, J.C. Tannehill, R.H. Pletcher, Computational Fluid Mechanics and Heat Transfer, McGraw Hill, 1997.

[14] C.S. Chen, A. Karageorghis, Y.S. Smyrlis, The Method of Fundamental Solutions: A Meshless Method, Dynamic Publishers, 2008.

[15] G.R. Liu, Y.T. Gu, An Introduction to Meshfree Methods and Their Programming, Springer, Netherlands, 2005.

[16] F. Williamson, Richard courant and the finite element method: a further look, Historia Mathematica 7 (1980) 369—378.

[17] Q.H. Qin, Y. Huang, BEM of postbuckling analysis of thin plates, Applied Mathematical Modelling 14 (1990) 544—548.

[18] Q.H. Qin, Y.-W. Mai, BEM for crack-hole problems in thermopiezoelectric materials, Engineering Fracture Mechanics 69 (2002) 577—588.

[19] J.T. Katsikadelis, The Boundary Element Method for Engineers and Scientists: Theory and Applications, Elsevier, 2016.

[20] R.A. Gingold, J.J. Monaghan, Smoothed particle hydrodynamics: theory and application to non-spherical stars, Monthly Notices of the Royal Astronomical Society 181 (1977) 375–389.

[21] E. Onate, S.R. Idelsohn, O.C. Zienkiewicz, R.L. Taylor, A finite point method in computational mechanics: applications to convective transport and fluid flow, International Journal for Numerical Methods in Engineering 39 (1996) 3839–3866.

[22] D.W. Kim, Y. Kim, Point collocation methods using the fast moving least-square reproducing kernel approximation, International Journal for Numerical Methods in Engineering 56 (2003) 1445–1464.

[23] E.J. Kansa, Multiquadrics-a scattered data approximation scheme with applications to computational fluid dynamics, Computers and Mathematics with Applications 19 (1990) 127–145.

[24] X. Zhang, X.H. Liu, K.Z. Song, M.W. Lu, Least-squares collocation meshless method, International Journal for Numerical Methods in Engineering 51 (2001) 1089–1100.

[25] B. Nayroles, G. Touzot, P. Villon, Generalizing the finite element method: diffuse approximation and diffuse elements, Computational Mechanics 10 (1992) 307–318.

[26] T. Belytschko, Y.Y. Lu, L. Gu, Element-free Galerkin methods, International Journal for Numerical Methods in Engineering 37 (1994) 229–256.

[27] W.K. Liu, S. Jun, Y.F. Zhang, Reproducing kernel particle methods, International Journal for Numerical Methods in Fluids 20 (1995) 1081–1106.

[28] J.M. Melenk, I. Babuska, The partition of unity finite element method: basic theory and applications, Computer Methods in Applied Mechanics and Engineering 139 (1996) 289–314.

[29] T.J. Liszka, C.A. Duarte, W.W. Tworzydlo, Hp-meshless cloud method, Computer Methods in Applied Mechanics and Engineering 139 (1996) 263–288.

[30] S.N. Atluri, S. Shen, The meshless local petrov-galerkin (MLPG) method: a simple & less-costly alternative to the finite element and boundary element methods, CMES-Computer Modeling in Engineering and Sciences 3 (2002) 11–51.

[31] G.R. Liu, Mesh Free Methods: Moving beyond the Finite Element Method, CRC Press, New York, 2003.

[32] J.M. Zhang, Z.H. Yao, H. Li, A hybrid boundary node method, International Journal for Numerical Methods in Engineering 53 (2002) 751–763.

[33] Y.T. Gu, G.R. Liu, A boundary point interpolation method for stress analysis of solids, Computational Mechanics 28 (2002) 47–54.

[34] G. Fairweather, A. Karageorghis, The method of fundamental solutions for elliptic boundary value problems, Advances in Computational Mathematics 9 (1998) 69–95.

[35] W. Chen, Meshfree boundary particle method applied to Helmholtz problems, Engineering Analysis with Boundary Elements 26 (2002) 577–581.

[36] W.Z. Chen, M. Tanaka, A meshless, integration-free, and boundary-only RBF technique, Computers and Mathematics with Applications 43 (2002) 379–391.

[37] E. Kita, N. Kamiya, Trefftz method: an overview, Advances in Engineering Software 24 (1995) 3–12.

[38] J.T. Chen, C.F. Wu, Y.T. Lee, K.H. Chen, On the equivalence of the Trefftz method and method of fundamental solutions for Laplace and biharmonic equations, Computers and Mathematics with Applications 53 (2007) 851–879.

[39] V.D. Kupradze, M.A. Aleksidze, The method of functional equations for the approximate solution of certain boundary value problems, USSR Computational Mathematics and Mathematical Physics 4 (1964) 82−126.

[40] A.H. Barnett, T. Betcke, Stability and convergence of the method of fundamental solutions for Helmholtz problems on analytic domains, Journal of Computational Physics 227 (2008) 7003−7026.

[41] X. Li, On convergence of the method of fundamental solutions for solving the Dirichlet problem of Poisson's equation, Advances in Computational Mathematics 23 (2005) 265−277.

[42] P. Mitic, Y.F. Rashed, Convergence and stability of the method of meshless fundamental solutions using an array of randomly distributed sources, Engineering Analysis with Boundary Elements 28 (2004) 143−153.

[43] Q.H. Qin, Green's Function and Boundary Elements of Multifield Materials, Elsevier, Oxford, 2007.

[44] Q.H. Qin, Green's functions of magnetoelectroelastic solids with a half-plane boundary or bimaterial interface, Philosophical Magazine Letters 84 (2004) 771−779.

[45] H. Wang, Q.H. Qin, Meshless approach for thermo-mechanical analysis of functionally graded materials, Engineering Analysis with Boundary Elements 32 (2008) 704−712.

[46] H. Wang, Q.H. Qin, A meshless method for generalized linear or nonlinear Poisson-type problems, Engineering Analysis with Boundary Elements 30 (2006) 515−521.

[47] H. Wang, Q.H. Qin, Some problems with the method of fundamental solution using radial basis functions, Acta Mechanica Solida Sinica 20 (2007) 21−29.

[48] H. Wang, Q.H. Qin, Y.L. Kang, A new meshless method for steady-state heat conduction problems in anisotropic and inhomogeneous media, Archive of Applied Mechanics 74 (2005) 563−579.

[49] H. Wang, Q.H. Qin, Y.L. Kang, A meshless model for transient heat conduction in functionally graded materials, Computational Mechanics 38 (2006) 51−60.

[50] L. Marin, D. Lesnic, The method of fundamental solutions for nonlinear functionally graded materials, International Journal of Solids and Structures 44 (2007) 6878−6890.

[51] C.J.S. Alves, On the choice of source points in the method of fundamental solutions, Engineering Analysis with Boundary Elements 33 (2009) 1348−1361.

[52] A. Karageorghis, A practical algorithm for determining the optimal pseudo-boundary in the method of fundamental solutions, Advances in Applied Mathematics and Mechanics 1 (2009) 510−528.

[53] C.M. Fan, C.S. Chen, J. Monroe, The method of fundamental solutions for solving convection-diffusion equations with variable coefficients, Advances in Applied Mathematics and Mechanics 1 (2009) 215−230.

[54] C.W. Chen, C.M. Fan, D.L. Young, K. Murugesan, C.C. Tsai, Eigenanalysis for membranes with stringers using the methods of fundamental solutions and domain decomposition, Computer Modeling in Engineering and Sciences 8 (2005) 29−44.

[55] C.S. Chen, H.A. Cho, M.A. Golberg, Some comments on the ill-conditioning of the method of fundamental solutions, Engineering Analysis with Boundary Elements 30 (2006) 405−410.

[56] T. Shigeta, D.L. Young, Method of fundamental solutions with optimal regularization techniques for the Cauchy problem of the Laplace equation with singular points, Journal of Computational Physics 228 (2009) 1903−1915.

[57] T. Wei, Y.C. Hon, L. Ling, Method of fundamental solutions with regularization techniques for Cauchy problems of elliptic operators, Engineering Analysis with Boundary Elements 31 (2007) 373–385.

[58] P.C. Hansen, Truncated singular value decomposition solutions to discrete ill-posed problems with ill-determined numerical rank, SIAM Journal on Scientific and Statistical Computing 11 (1990) 503–518.

[59] P.A. Ramachandran, Method of fundamental solutions: singular value decomposition analysis, Communications in Numerical Methods in Engineering 18 (2002) 789–801.

[60] S.A. Teukolsky, W.T. Vetterling, B.P. Flannery, Numerical Recipes in C: The Art of Scientific Computing, Cambridge University Press, New York, 1992.

[61] M.A. Golberg, The method of fundamental solutions for Poisson's equation, Engineering Analysis with Boundary Elements 16 (1995) 205–213.

[62] M.A. Golberg, C.S. Chen, The Method of Fundamental Solution for Potential, Helmholtz and Diffusion Problems, Computational Mechanics Publications, Southampton, 1998.

[63] Z.W. Zhang, H. Wang, Q.H. Qin, Method of fundamental solutions for nonlinear skin bioheat model, Journal of Mechanics in Medicine and Biology 14 (2014) 1450060.

[64] Z.W. Zhang, H. Wang, Q.H. Qin, Meshless method with operator splitting technique for transient nonlinear bioheat transfer in two-dimensional skin tissues, International Journal of Molecular Sciences 16 (2015) 2001–2019.

Chapter 2

Mechanics of solids and structures

Chapter outline

2.1 Introduction

Solid mechanics is one of the important branches of physical science concerned with the deformation and motion of continuous solid media under applied external loadings such as forces, displacements, and accelerations that result in inertial force in the bodies, thermal changes, chemical interactions, electromagnetic forces, and so on. In the context of continuum solid mechanics [1,2], the basic theory is generally built based on two foundations: (1) the basic laws of motion describing the equilibrium of a continuum body under external loadings and induced interior forces. They are valid for all continuum bodies; (2) a constitutive theory describing the mechanical behavior of materials used for the construction of a continuum body. The resulting equations contain some material parameters that can be determined through experiment.

Methods of Fundamental Solutions in Solid Mechanics. https://doi.org/10.1016/B978-0-12-818283-3.00002-6

The purpose of this chapter is to provide the necessary mathematical equations governing the linear responses of elastic solids to external changes. The linear elastic solids here include general three-dimensional solids, reduced two-dimensional solids, simplified beam structures and thin plate structures, and extensive two-dimensional piezoelectric solids. These equations form the basis of the present meshless collocation methods. More details of these equations can be found in most textbooks on elasticity, plate, and piezoelectricity, i.e., [3−6].

2.2 Basic physical quantities

2.2.1 Displacement components

For a three-dimensional (3D) elasticity problem, the deformation at any point $\mathbf{x}(x, y, z)$, e.g., the point P in Fig. 2.1, in a solid can be measured by the following three displacement components (m) that are functions of the spatial coordinate x, y, and z,

$$\mathbf{u}(\mathbf{x}) = \{\, u(x, y, z) \quad v(x, y, z) \quad w(x, y, z) \,\}^{\mathrm{T}} \tag{2.1}$$

2.2.2 Stress components

Stress ($\mathrm{N/m^2}$) is a very important concept in solid mechanics, representing the interior interaction in continuum media due to applied external forces. To illustrate this concept, we consider a 3D elastic solid, as shown in Fig. 2.2A. It is assumed that the solid is supported at certain locations and is subjected to various external forces that may be distributed over the volume or on the boundary, or are concentrated at several locations.

For such a solid, internal forces may exist between the parts of the body. To illustrate the magnitude of these internal forces at any point, e.g., point O in the body, we can imagine that the body is divided into two parts, A and B, by a

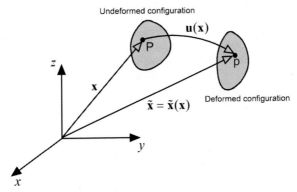

FIGURE 2.1 Motion and deformation of a point in a continuum.

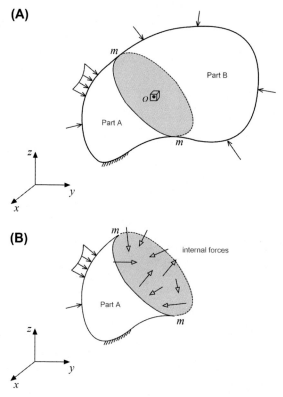

FIGURE 2.2   (A) A continuum of solid (B) internal forces distributed over the cross-section *mm*.

cross-section *mm* through this point. Considering one of these parts, for instance the part A, it can be stated that it is in equilibrium under the action of external forces, and the internal forces distributed over the cross-section *mm* represent the actions of the material of the part B on the material of the part A, as indicated in Fig. 2.2B. If these internal forces are continuously distributed over the surface *mm* in a uniform or nonuniform manner, their magnitudes can be defined by their intensity, i.e., the amount of force per unit area. We usually call this intensity **stress** [4].

    To characterize the stress acting at the point *O*, we can cut out a very small cubic element whose center is at point *O*, as shown in Fig. 2.2A, with three pairs of opposite faces parallel to the coordinate axes. If we use the letter σ to represent the normal stress perpendicular to the area on which it acts and the letter τ for the shearing stress acting in the plane of the area, the stress components acting on each area of this element can be shown as in Fig. 2.3. The subscript in the normal stress diagram indicates that the stress is acting on a plane normal to the axis of the subscript letter. The normal stress is taken to be positive when it produces tension and negative when it produces

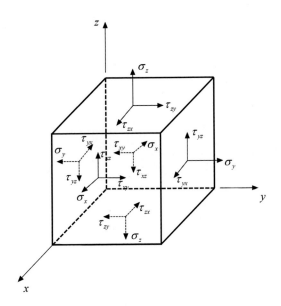

FIGURE 2.3 Positive stress components on a small cubic element.

compression. For the shearing stress, there are two subscript letters. The first letter indicates the direction of the normal to the plane under consideration, and the second letter indicates the direction of the component of the shearing stress. Positive directions of the components of shearing stress on any surface of the cubic element are taken as the positive directions of the coordinate axes if a tensile stress on the same surface would have the positive direction of the corresponding axis. If the tensile stress has a direction opposite to the positive axis, the positive direction of the shearing stress components should be reversed. Following this rule, all stress components on each surface of the cubic element are positive, as displayed in Fig. 2.3.

From this discussion, we know that there are three normal stresses, σ_x, σ_y, and σ_z, and six shearing stresses, τ_{xy}, τ_{xz}, τ_{yx}, τ_{yz}, τ_{zx}, and τ_{zy}, acting on the six surfaces of the cubic element. By simply considering the equilibrium of the element, we have

$$\tau_{xy} = \tau_{yx}, \tau_{yz} = \tau_{zy}, \tau_{xz} = \tau_{zx} \tag{2.2}$$

so the number of shearing stresses can be reduced to three, for instance, τ_{xy}, τ_{yz}, and τ_{zx}.

The six stress components are therefore sufficient to describe the stresses acting on the coordinate planes through a point, and they are called the components of stress at the point. In vector form, the stress can be written as

$$\boldsymbol{\sigma} = \left\{ \sigma_x \ \sigma_y \ \sigma_z \ \tau_{yz} \ \tau_{xz} \ \tau_{xy} \right\}^{\mathrm{T}} \tag{2.3}$$

2.2.3 Strain components

Strain components (m/m) are introduced to measure the deformation of solids. For 3D cases, there are six strain components at a point in a solid, corresponding to the six stress components. They can be written in vector form as

$$\boldsymbol{\varepsilon} = \left\{ \varepsilon_x \ \varepsilon_y \ \varepsilon_z \ \gamma_{yz} \ \gamma_{xz} \ \gamma_{xy} \right\}^{\mathrm{T}} \tag{2.4}$$

where ε_x, ε_y, and ε_z are normal strains measuring changes in length along a specific direction, and γ_{yz}, γ_{xz}, and γ_{xy} are engineering shear strains measuring changes in angles with respect to two specific directions. For example, in a general deformable solid, as shown in Fig. 2.4, the applied external forces and the specific boundary displacement constraints can cause deformation of the solid. As a result, the line element connecting arbitrary points P and Q taken in the nondeformed solid becomes $P'Q'$ after the deformation that occurs with changes in length and angle [4].

2.3 Equations for three-dimensional solids

2.3.1 Strain-displacement relation

Theoretically, both normal strain and shear strain can be regarded as a rate of displacement variation and angle per unit length. Thus, the components of strain can be obtained by derivatives of the displacements for small deformation in solids. The strain-displacement relation can be written in matrix form as follows [3, 4]:

$$\boldsymbol{\varepsilon} = \mathbf{L}\mathbf{u} \tag{2.5}$$

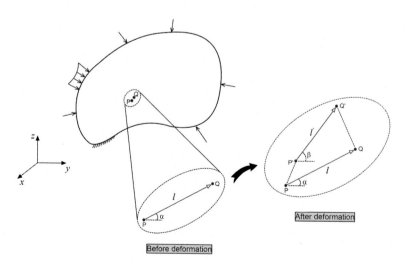

FIGURE 2.4   Illustration of strain concept.

where

$$\mathbf{u} = \{ u \quad v \quad w \}^{\mathrm{T}} \tag{2.6}$$

is a displacement vector consisting of three displacement components u, v, and w in the x, y, and $z$directions, respectively. $\mathbf{L}$ is a matrix differential operator given by

$$\mathbf{L} = \begin{bmatrix} \dfrac{\partial}{\partial x} & 0 & 0 \\[2mm] 0 & \dfrac{\partial}{\partial y} & 0 \\[2mm] 0 & 0 & \dfrac{\partial}{\partial z} \\[2mm] 0 & \dfrac{\partial}{\partial z} & \dfrac{\partial}{\partial y} \\[2mm] \dfrac{\partial}{\partial z} & 0 & \dfrac{\partial}{\partial x} \\[2mm] \dfrac{\partial}{\partial y} & \dfrac{\partial}{\partial x} & 0 \end{bmatrix} \tag{2.7}$$

Eq. (2.5) describes the relation of strain and displacement. Thus, in practice, it is also known as the strain-displacement relation.

In addition to the strain-displacement relation, the basic equations describing 3D solid deformation are the equilibrium equations, the constitutive equations, and the boundary conditions [4].

2.3.2 Equilibrium equations

The equilibrium equations describe the relationship between the stress and an external force and can be derived by considering the balance of a small solid element, as indicated in Fig. 2.5.

It is assumed that ρ is the mass density of the material (kg/m^3), c is the damping coefficient (kg/s/m^3), and

$$\mathbf{b} = \{ b_x \quad b_y \quad b_z \}^{\mathrm{T}} \tag{2.8}$$

is the body force per unit volume (N/m^3). By considering the force balance of the microelement along the x, y, and z directions, we finally have

$$\mathbf{L}^{\mathrm{T}}\boldsymbol{\sigma} + \mathbf{b} = \rho \ddot{\mathbf{u}} + c\dot{\mathbf{u}} \tag{2.9}$$

where $\ddot{\mathbf{u}} = \partial^2 \mathbf{u}/\partial t^2$ and $\dot{\mathbf{u}} = \partial \mathbf{u}/\partial t$ denote acceleration vector and velocity vector, respectively, at any time instance t.

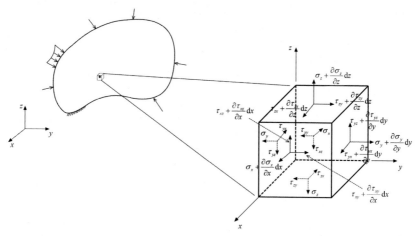

FIGURE 2.5   Equilibrium of solid microelement.

For static problems in particular, the equilibrium equations can be simplified as

$$\mathbf{L}^{\mathrm{T}}\boldsymbol{\sigma} + \mathbf{b} = 0 \tag{2.10}$$

2.3.3 Constitutive equations

The constitutive equation, sometimes called the generalized Hooke's law, gives the relationship between the stress and the strain in a given deformed solid. The generalized Hooke's law for general anisotropic elastic materials can be given in matrix form as shown:

$$\boldsymbol{\sigma} = \mathbf{D}\boldsymbol{\varepsilon} \tag{2.11}$$

where $\mathbf{D}$ is a symmetric matrix dependent on the material, which must be determined through standard experimental tests, and can be written as follows [2,7]:

$$\mathbf{D} = \begin{bmatrix} D_{11} & D_{12} & D_{13} & D_{14} & D_{15} & D_{16} \\ & D_{22} & D_{23} & D_{24} & D_{25} & D_{26} \\ & & D_{33} & D_{34} & D_{35} & D_{36} \\ & & & D_{44} & D_{45} & D_{46} \\ & \text{SYM} & & & D_{55} & D_{56} \\ & & & & & D_{66} \end{bmatrix} \tag{2.12}$$

Because the matrix $\mathbf{D}$ remains symmetric, only 21 components (N/m^2 or Pa) are actually independent. Besides, the matrix $\mathbf{D}$ is often called the stiffness matrix, so each of the components is an elastic stiffness constant. The term "stiffness" here is used to measure how "hard" this solid is. A large elastic stiffness constant means that a larger force is needed to deform this solid. In other words, the solid is "hard."

Specially, for isotropic elastic material, which is the simplest type of material, $\mathbf{D}$ can be reduced to

$$\mathbf{D} = \begin{bmatrix} D_{11} & D_{12} & D_{12} & 0 & 0 & 0 \\ & D_{11} & D_{12} & 0 & 0 & 0 \\ & & D_{11} & 0 & 0 & 0 \\ & & & D_{44} & 0 & 0 \\ & \text{SYM} & & & D_{44} & 0 \\ & & & & & D_{44} \end{bmatrix} \tag{2.13}$$

where

$$D_{11} = \frac{E(1 - \nu)}{(1 - 2\nu)(1 + \nu)}$$

$$D_{12} = \frac{E\nu}{(1 - 2\nu)(1 + \nu)} \tag{2.14}$$

$$D_{44} = \frac{D_{11} - D_{12}}{2} = \frac{E}{2(1 + \nu)} = G$$

in which E, ν, and G are Young's modulus (N/m^2 or Pa), Poisson's ratio, and shear modulus (N/m^2 or Pa) of the material, respectively.

Finally, substituting Eq. (2.5) into the constitutive relations (2.11), and then into the equilibrium Eq. (2.9), we have the following set of partial differential equations in terms of the primary dependent variable $\mathbf{u}$

$$\mathbf{L}^{\mathrm{T}}\mathbf{D}\mathbf{L}\mathbf{u} + \mathbf{b} = \rho\ddot{\mathbf{u}} + c\dot{\mathbf{u}} \tag{2.15}$$

for dynamic problems or

$$\mathbf{L}^{\mathrm{T}}\mathbf{D}\mathbf{L}\mathbf{u} + \mathbf{b} = 0 \tag{2.16}$$

for static problems.

2.3.4 Boundary conditions

For a complete static boundary value problem, the boundary conditions should be provided to combine with the governing equations, whereas for dynamic problems, the initial conditions must be given together with the boundary conditions. Here, we consider static problems only.

For a typical static 3D linear elastic problem, the necessary boundary conditions include:

1. Displacement boundary conditions

$$\mathbf{u} = \bar{\mathbf{u}} \tag{2.17}$$

where $\bar{\mathbf{u}} = \left\{ \bar{u}_x \quad \bar{u}_y \quad \bar{u}_z \right\}^{\mathrm{T}}$ denotes the prescribed displacement value on the displacement boundary part.

2. Traction boundary conditions

$$\mathbf{t} = \bar{\mathbf{t}} \tag{2.18}$$

where $\mathbf{t} = \left\{ t_x \quad t_y \quad t_z \right\}^{\mathrm{T}}$ and $\bar{\mathbf{t}} = \left\{ \bar{t}_x \quad \bar{t}_y \quad \bar{t}_z \right\}^{\mathrm{T}}$ are the traction vector and the prescribed traction value, respectively, on the traction boundary part. The traction vector can be expressed by

$$\mathbf{t} = \mathbf{A}\boldsymbol{\sigma} \tag{2.19}$$

with

$$\mathbf{A} = \begin{bmatrix} n_1 & 0 & 0 & 0 & n_3 & n_2 \\ 0 & n_2 & 0 & n_3 & 0 & n_1 \\ 0 & 0 & n_3 & n_2 & n_1 & 0 \end{bmatrix} \tag{2.20}$$

and n_i $(i = 1,2,3)$ is the component of the unit outward normal vector to the boundary.

It should be noted that in some cases, The displacement and its conjugate force cannot be prescribed simultaneously. The displacement and traction boundary conditions may be given together, but the total number of boundary conditions on the boundary should be three for the 3D case. For example, for the boundary part perpendicular to the x-axis shown in Fig. 2.6, the simple support along the x-direction constrains the displacement along the x-direction, while the tractions along the y- and z-directions acting on the plane remain free. Therefore, the given boundary conditions for such a case can be written as

$$u_x = 0$$
$$t_y = 0 \tag{2.21}$$
$$t_z = 0$$

Theoretically, the equations described earlier are applicable to any 3D linear elastic solids. However, treating all the structural components as a 3D solid makes analysis very complex for some special cases. As we know, some

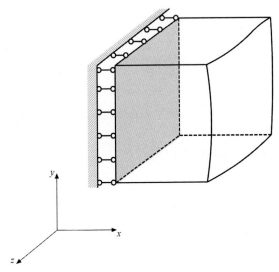

FIGURE 2.6 Illustration of boundary conditions consisting of displacement and traction constraints.

structural types can be simplified due to their geometric and loading features. For example, a beam is slim and long compared to its cross-section, and the main deformation is bending. Thus, simplified beam bending theory may be more appropriate than the full 3D linear elastic theory for engineering applications and can dramatically reduce analytical and computational effort. Certainly, all simplified theories should be derived from the full 3D linear elastic theory. In the following sections, some simplified theories are given.

2.4 Equations for plane solids

2.4.1 Plane stress and plane strain

When 3D solids meet certain conditions, they can be reduced to 2D ones. For example, if a solid is very thin along the z-direction compared to its dimensions in the xy plane, such as the thin plate in Fig. 2.7, and it is subjected to external loads applied in the xy plane, the stresses σ_z, τ_{xz}, and τ_{yz} can be ignored, that is,

$$\sigma_z = 0, \quad \tau_{xz} = 0, \quad \tau_{yz} = 0 \tag{2.22}$$

In such cases, the 3D elasticity can be reduced to 2D elasticity, and this case is usually called plane stress.

Another 2D case is plane strain, as illustrated in Fig. 2.8. In such cases, when the thickness of a solid in the z-direction is very great compared to the

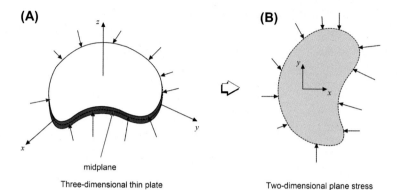

FIGURE 2.7   Configurations of plane stress.

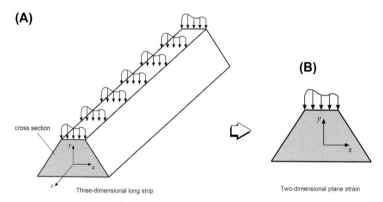

FIGURE 2.8   Configurations of plane strain.

dimensions in the x- and y- directions and all external forces are applied uniformly along the z-direction, the strain components in the z-direction can be assumed to be zero, that is,

$$\varepsilon_z = 0, \quad \gamma_{xz} = 0, \quad \gamma_{yz} = 0 \tag{2.23}$$

For both plane stress and plane strain problems, the solving space is reduced to the xy plane. Hence the basic unknown fields such as displacements, strains, and stresses become functions of x and y only, that is,

$$\mathbf{u} = \{ u(x, y) \quad v(x, y) \}^{\mathrm{T}} \tag{2.24}$$

$$\boldsymbol{\sigma} = \{ \sigma_x(x, y) \quad \sigma_y(x, y) \quad \tau_{xy}(x, y) \}^{\mathrm{T}} \tag{2.25}$$

$$\boldsymbol{\varepsilon} = \{ \varepsilon_x(x, y) \quad \varepsilon_y(x, y) \quad \gamma_{xy}(x, y) \}^{\mathrm{T}} \tag{2.26}$$

2.4.2 Governing equations

Unknown displacement, strain, and stress fields are connected by three field equations: strain-displacement relation, constitutive relation, and equilibrium equations:

$$\boldsymbol{\varepsilon} = \mathbf{L}\mathbf{u} \tag{2.27}$$

$$\boldsymbol{\sigma} = \mathbf{D}\boldsymbol{\varepsilon} \tag{2.28}$$

$$\mathbf{L}^{\mathrm{T}}\boldsymbol{\sigma} + \mathbf{b} = 0 \tag{2.29}$$

where

$$\mathbf{L} = \begin{bmatrix} \dfrac{\partial}{\partial x} & 0 \\[2ex] 0 & \dfrac{\partial}{\partial y} \\[2ex] \dfrac{\partial}{\partial y} & \dfrac{\partial}{\partial x} \end{bmatrix} \tag{2.30}$$

and

$$\mathbf{b} = \{\, b_x \quad b_y \,\}^{\mathrm{T}} \tag{2.31}$$

In Eq. (2.28), the material matrix $\mathbf{D}$ has different forms for plane stress and plane strain states. For example, for isotropic material, we have

$$\mathbf{D} = \frac{E}{(1+v)(1-2v)} \begin{bmatrix} 1-v & v & 0 \\ v & 1-v & 0 \\ 0 & 0 & 0.5-v \end{bmatrix} \quad \text{(plane strain)} \tag{2.32}$$

and

$$\mathbf{D} = \frac{E}{1-v^2} \begin{bmatrix} 1 & v & 0 \\ v & 1 & 0 \\ 0 & 0 & \dfrac{1-v}{2} \end{bmatrix} \quad \text{(plane stress)} \tag{2.33}$$

It is necessary to point out that, from the 3D constitutive equations given in Section 2.3, we can observe that, for the plane strain state, the normal stress σ_z is generally nonzero, but it does not enter the governing equations as a basic unknown. Similarly, for the plane stress state, the normal strain ε_z is generally nonzero, but it too does not enter the governing equations as a basic unknown.

2.4.3 Boundary conditions

Boundary conditions prescribed on the boundary of a computational domain can be divided into two types: displacement boundary conditions or traction boundary conditions, over which displacements and tractions, respectively, are specified.

Displacement boundary conditions are prescribed on the displacement boundary part Γ_u in the form

$$\mathbf{u} = \bar{\mathbf{u}} \tag{2.34}$$

where $\bar{\mathbf{u}} = \{ \bar{u}_x \quad \bar{u}_y \}^T$ denotes the prescribed displacement value.

Traction boundary conditions are prescribed on the traction boundary part Γ_t and take the form

$$\mathbf{t} = \bar{\mathbf{t}} \tag{2.35}$$

where $\mathbf{t} = \{ t_x \quad t_y \}^T$ and $\bar{\mathbf{t}} = \{ \bar{t}_x \quad \bar{t}_y \}^T$ are the traction vector and the prescribed traction value, respectively, which are defined as force per unit area.

Generally, the traction vector connects to the stresses by the expression

$$\mathbf{t} = \mathbf{A}\boldsymbol{\sigma} \tag{2.36}$$

where

$$\mathbf{A} = \begin{bmatrix} n_1 & 0 & n_2 \\ 0 & n_2 & n_1 \end{bmatrix} \tag{2.37}$$

and n_i (i = 1,2) is the component of the unit outward normal vector to the boundary.

2.5 Equations for Euler–Bernoulli beams

In various engineering structures, there is a type of structural element that has one of its dimensions that is much larger than the other two and primarily resists transverse loads. Such a structural element is usually called a beam. Generally, the axis of a beam is defined along the direction with the longest dimension, and a cross-section normal to this axis is assumed to remain uniform or vary smoothly along the length of the beam. For such structures, the corresponding elastic theory describing their deformable behaviors is commonly called "beam theory," which plays an important role in structural analysis because it provides a simple and effective tool for engineers to analyze numerous beam structures.

To describe the Euler–Bernoulli beam bending theory (also known as classical beam theory) conveniently, we establish the following Cartesian coordinate system in Fig. 2.9. The x-coordinate is taken along the length of the beam and remains unstretchable during deformation, the z-coordinate is along

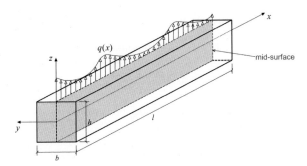

FIGURE 2.9   Euler–Bernoulli beam model subjected to transverse load.

the thickness (or height) of the beam, and the y-coordinate is taken along the width of the beam (see the coordinate system in Fig. 2.9 for a rectangular cross-section beam). We commonly define the x-axis as the neutral axis and take the z-axis as the symmetric axis of the cross-section. Thus, the longitudinal plane of symmetry is in the xoz plane, which is also called the plane of bending. Recall that, depending on the cross-sectional shape of the beam and its composition, the neutral axis may not be located at the center of the beam height. In Fig. 2.9, the beam is subjected to a distributed transverse load q(x) in the xoz plane (midsurface of the beam). The transverse distributed loading q(x) is defined in units of force per length.

2.5.1 Deformation mode

Euler–Bernoulli beam theory is established on the following three primary assumptions [6]:

- The shape and geometry of cross-sections of the beam do not change in a significant manner und er applied transverse loads. This means that a cross-section can be assumed as a rigid surface during deformation and can only rotate.
- During deformation, the cross-section of the beam is assumed to remain planar and normal to the deformed axis of the beam, as shown in Fig. 2.10.
- Although the neutral axis of the beam becomes curved after deformation, the deformed angles (slopes) are small.

Based on the Euler-Bernoulli assumptions given before, the displacement fields can be derived (see Fig. 2.11):

$$u(x, y, z) = u_0(x) - z \sin \theta$$
$$v(x, y, z) = 0 \tag{2.38}$$
$$w(x, y, z) = w_0(x) + z \cos \theta - z$$

where u, v, and w are the displacements of an arbitrary point (x, y, z) along the x-, y-, and z-directions, respectively, and u_0 and w_0 are the displacements

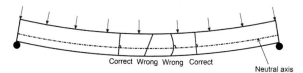

FIGURE 2.10 Assumption of planar cross-section in thin beam.

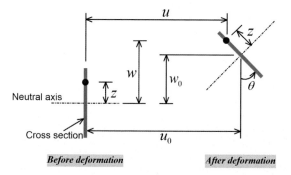

FIGURE 2.11 Kinematics of deformation of Euler-Bernoulli beam.

of the point $(x, 0, 0)$. θ denotes the rotation of the cross-section (positive if it is counterclockwise).

Under the assumption of a small deformation, we have

$$\sin\theta \approx \theta, \quad \cos\theta \approx 1 \tag{2.39}$$

So the displacement fields can be simplified as

$$
\begin{aligned}
u(x, y, z) &= u_0(x) - z\theta \\
v(x, y, z) &= 0 \\
w(x, y, z) &= w_0(x)
\end{aligned}
\tag{2.40}
$$

Simultaneously, based on the Euler–Bernoulli assumptions, the rotation θ of the cross-section can be expressed as

$$\theta = \frac{dw}{dx} \tag{2.41}$$

Therefore, the displacement fields for Euler-Bernoulli beam theory can be written as

$$
\begin{aligned}
u(x, y, z) &= u_0(x) - z\frac{dw(x)}{dx} \\
v(x, y, z) &= 0 \\
w(x, y, z) &= w_0(x)
\end{aligned}
\tag{2.42}
$$

2.5.2 Governing equations

With the strain-displacement relation, strain components can be naturally derived by

$$\varepsilon_x = \frac{\partial u}{\partial x} = \frac{du_0}{dx} - z\frac{d^2 w}{dx^2}$$

$$\gamma_{xz} = 2\varepsilon_{xz} = \frac{\partial u}{\partial z} + \frac{\partial w}{\partial x} = 0 \tag{2.43}$$

and other components are zero

where γ_{xz} is the engineering shear strain.

The one-dimensional stress-strain relationship for a linear elastic isotropic beam can be written as

$$\sigma_x = E\varepsilon_x \tag{2.44}$$

where E is the Young's modulus of the material. Here, we allow the Young's modulus to vary freely along the axis of the beam (this may occur, for example, in functionally graded materials), and the cross-section is homogeneous. Hence, in this case, the Young's modulus is a function of the axis coordinate x, that is, $E = E(x)$.

Replacing the strain field for this case gives

$$\sigma_x = E\left(\frac{du_0}{dx} - z\frac{d^2 w}{dx^2}\right) \tag{2.45}$$

Stress resultants are equivalent force systems that represent the integral effect of the internal stresses acting on the cross-section of the beam so that they eliminate the dependency of stresses on the spatial coordinates y and z of the cross-section. Here, for the Euler–Bernoulli beam under consideration, the resultant axial force N, bending moment M, and shear force Q can be defined by the expressions

$$N = \int_A \sigma_x dA$$

$$M = \int_A z\sigma_x dA \tag{2.46}$$

$$Q = \int_A \tau_{xz} dA$$

Note that positive stresses produce positive force resultants. Thus Fig. 2.12 uses positive sign conventions for N, M, and Q.

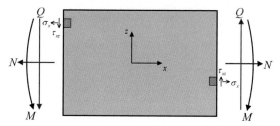

FIGURE 2.12 Positive sign conventions for axial force, moment, and shear.

Replacing the stress component in (2.46), we have

$$N = EA \frac{du_0}{dx}$$

$$M = E \frac{du_0}{dx} \int_A z \, dA - E \frac{d^2 w_0}{dx^2} \int_A z^2 \, dA = -EI_y \frac{d^2 w_0}{dx^2} \tag{2.47}$$

$$Q = \int_A \tau_{xz} \, dA = 0$$

where A is the cross-sectional area, and $I_y = \int_A z^2 \, dA$ is the moment of inertia or the second moment of area of the cross-section about the y-axis. Note that the y-axis is taken through the geometric centroid of the cross-section, so the integral $\int_A z \, dA = 0$. Therefore, the neutral axis in the Euler–Bernoulli beam bending must go through the geometric centroid of each cross-section.

Although here we obtain zero transverse shear forces along the cross-section, that is to say that the effect of shear forces on beam deformations is neglected in accordance with the fundamental assumptions, we can recover transverse shear forces by considering the equilibrium of a microelement of length dx, as illustrated in Fig. 2.13.

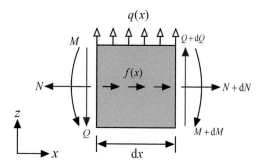

FIGURE 2.13 Equilibrium of a beam element subjected to axial and transverse loads.

Considering the force equilibriums in the vertical z-direction and the horizontal x-direction on the beam microelement, we have

$$\sum F_x = 0: \quad \frac{\mathrm{d}}{\mathrm{d}x} N(x) + f(x) = 0$$

$$\sum F_z = 0: \quad \frac{\mathrm{d}}{\mathrm{d}x} Q(x) + q(x) = 0 \tag{2.48}$$

Additionally, summing the moments about the y-axis through the center of a beam microelement yields the moment equilibrium equation:

$$\sum M_y = 0: \quad \frac{\mathrm{d}}{\mathrm{d}x} M(x) - Q(x) = 0 \tag{2.49}$$

Substituting the constitutive relation (2.47) into the two governing differential equations in Eq. (2.48), we obtain the equations:

$$\frac{\mathrm{d}}{\mathrm{d}x} \left(EA \frac{\mathrm{d}u_0}{\mathrm{d}x} \right) + f(x) = 0 \tag{2.50}$$

$$\frac{\mathrm{d}^2}{\mathrm{d}x^2} \left(EI_y \frac{\mathrm{d}^2 w_0}{\mathrm{d}x^2} \right) = q(x) \tag{2.51}$$

Alternatively, expanding Eq. (2.51) produces

$$\frac{\mathrm{d}^2 EI_y}{\mathrm{d}x^2} \frac{\mathrm{d}^2 w_0}{\mathrm{d}x^2} + 2 \frac{\mathrm{d}EI_y}{\mathrm{d}x} \frac{\mathrm{d}^3 w_0}{\mathrm{d}x^3} + EI_y \frac{\mathrm{d}^4 w_0}{\mathrm{d}x^4} = q(x) \tag{2.52}$$

which is the governing differential equation of the Euler-Bernoulli beam describing the relationship between the beam's deflection and the applied transverse load.

At the same time, the recovered transverse shear force Q is recomputed from Eq. (2.49) as

$$Q = \frac{\mathrm{d}M}{\mathrm{d}x} = \frac{\mathrm{d}}{\mathrm{d}x} \left(-EI_y \frac{\mathrm{d}^2 w_0}{\mathrm{d}x^2} \right) = -\frac{\mathrm{d}(EI_y)}{\mathrm{d}x} \frac{\mathrm{d}^2 w_0}{\mathrm{d}x^2} - EI_y \frac{\mathrm{d}^3 w_0}{\mathrm{d}x^3} \tag{2.53}$$

In particular, if the material constant E and the moment of inertia I_y are constants, the governing Eq. (2.52) can be simplified as

$$EI_y \frac{\mathrm{d}^4 w_0}{\mathrm{d}x^4} = q(x) \tag{2.54}$$

and

$$Q = -EI_y \frac{\mathrm{d}^3 w_0}{\mathrm{d}x^3} \tag{2.55}$$

In summary, for convenience, the symbol w_0 in the governing Eqs. (2.52) and (2.54) is replaced with w, which is the four times differentiable sought solution. Thus, in the case of a changed cross-section that causes I_y to be a function of the spatial variable x, Eq. (2.52) is rewritten as

$$\frac{d^2 EI_y}{dx^2}\frac{d^2 w}{dx^2} + 2\frac{dEI_y}{dx}\frac{d^3 w}{dx^3} + EI_y\frac{d^4 w}{dx^4} = q(x) \qquad (2.56)$$

with

$$Q = -\frac{dEI_y}{dx}\frac{d^2 w}{dx^2} - EI_y\frac{d^3 w}{dx^3} \qquad (2.57)$$

$$M = -EI_y\frac{d^2 w}{dx^2} \qquad (2.58)$$

Typically, for a uniform beam, the governing Eq. (2.54) is rewritten as

$$EI_y\frac{d^4 w}{dx^4} = q(x) \qquad (2.59)$$

Simultaneously, the internal resultant force on the cross-section can be given by

$$Q = -EI_y\frac{d^3 w}{dx^3} \qquad (2.60)$$

$$M = -EI_y\frac{d^2 w}{dx^2} \qquad (2.61)$$

2.5.3 Boundary conditions

Finally, the relevant boundary conditions must be added to make the system complete and solvable. For the beam bending problem, which is governed by the fourth-order linear inhomogeneous Eq. (2.59), there are four boundary conditions, with two at each end. The essential boundary conditions are on the variables w and θ, whereas the natural boundary conditions are on the variables Q and M. The boundary condition sets on w and Q, θ and M are disjoint; that is, if w is prescribed, Q cannot be prescribed, and vice versa. So, we can depict the boundary conditions as follows:

At any point, there are two sets:

Set 1: w is specified, or Q is specified.
Set 2: θ is specified, or M is specified.

For example, Fig. 2.14 shows some typical constraints widely used in engineering, namely the fixed end, the pinned support, and the free end, and the corresponding boundary conditions meeting the aforementioned specifications are given in the figure.

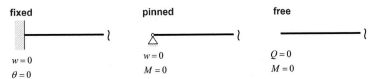

FIGURE 2.14 Typical static boundary conditions for beams.

2.5.4 Continuity requirements

From beam bending theory, it is revealed that a sudden change in the beam loading can produce discontinuous internal force solutions. However, deformation discontinuities including deflection and slope discontinuities are not permitted in beams.

In the mechanics of continuum solids, the discontinuity of a given function f at x_0 can be defined by a square bracket enclosure

$$[f(x_0)] = f(x_0^+) - f(x_0^-) \tag{2.62}$$

where x_0^+ and x_0^- denote the values of the argument on the right and left side, respectively, of a discontinuity.

The quasistatic beam theory generally requires continuity of deformation and rotation, as shown in Fig. 2.15

$$[w] = 0, \quad [\theta] = 0 \tag{2.63}$$

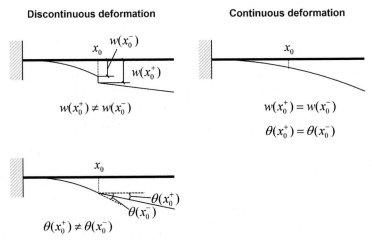

FIGURE 2.15 Illustration of deflection and slope discontinuities that are not permitted in beam deformation.

2.6 Equations for thin plates

Thin plates are another special and important type of engineering structure, in which the transverse dimension or thickness of the plate is significantly small compared to the length and width dimensions [8, 9]. Generally, it is assumed that the thickness of thin plate is in the approximate range of 1/20−1/100 of its span.

   In this section, the basic equations that describe the bending behavior of thin plates are derived, based on generalization of the one-dimensional Euler−Bernoulli beam theory, which exploits the slender shape of a beam. To establish the 2D thin plate bending theory, also called the Kirchhoff theory, a Cartesian coordinate system (x, y, z) is employed, as shown in Fig. 2.16. The top and bottom surfaces lie at $z = \pm t/2$, and the flat surface $z = 0$ is the plate midsurface, which provides a convenient reference plane for the derivation of the governing equations of the thin plate. Here, t represents the plate's thickness. Under the moment and shear loads applied to the plate's edges, and the transverse distributed loading in the z-direction, the plate's bending behavior can be observed.

2.6.1 Deformation mode

Like the Euler−Bernoulli beam theory, the bending deformation theory of a thin plate is established based on three primary assumptions [6,9,10]:

- Line elements perpendicular to the middle surface of the plate before deformation remain normal and unstretched after deformation.
- The deflections of the plate are small in comparison to its thickness, so the linear strain-displacement relations are valid.

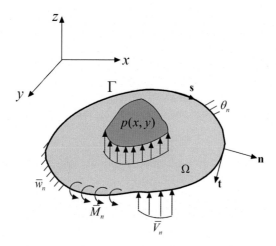

FIGURE 2.16   Configurations of a thin plate under arbitrary transverse loads.

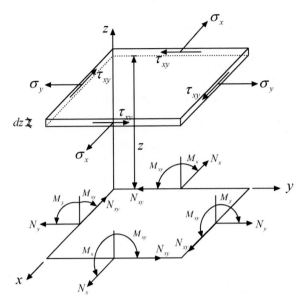

FIGURE 2.17   Positive directions of membrane stress components and their resultants through the thickness.

- The normal stress in the thickness direction can be neglected because the thickness is small.

Based on these assumptions, the displacement fields can be assumed by

$$u(x, y, z) = u_0(x, y) - z\theta_x$$
$$v(x, y, z) = v_0(x, y) - z\theta_y \qquad (2.64)$$
$$w(x, y, z) = w_0(x, y)$$

where u_0, v_0, and w_0 are the displacements of the midsurface, and the coefficients of the linear terms θ_x and θ_y can be interpreted as rotations of the midsurface in the x- and y-directions.

Under the assumption of small deformation, the rotations of the midsurface can be written in terms of the first-order derivative of the normal displacement w_0, which represents the rotations of the unit normal to the midsurface, that is,

$$\theta_x = \frac{\partial w_0}{\partial x}, \qquad \theta_y = \frac{\partial w_0}{\partial y} \qquad (2.65)$$

Thus, Eq. (2.64) can be further written as

$$u(x, y, z) = u_0(x, y) - z\frac{\partial w_0}{\partial x}$$

$$v(x, y, z) = v_0(x, y) - z\frac{\partial w_0}{\partial y} \qquad (2.66)$$

$$w(x, y, z) = w_0(x, y)$$

2.6.2 Governing equations

Substituting Eq. (2.66) into the strain-displacement relation yields the in-plane strain components

$$\varepsilon_x = \frac{\partial u}{\partial x} = \varepsilon_{x0}(x, y) + z\kappa_x(x, y)$$

$$\varepsilon_y = \frac{\partial v}{\partial y} = \varepsilon_{y0}(x, y) + z\kappa_y(x, y) \qquad (2.67)$$

$$\gamma_{xy} = \frac{\partial u}{\partial y} + \frac{\partial v}{\partial x} = \gamma_{xy0}(x, y) + 2z\kappa_{xy}(x, y)$$

with all other components being zero,

where

$$\varepsilon_{x0} = \frac{\partial u_0}{\partial x}, \quad \kappa_x = -\frac{\partial^2 w}{\partial x^2}$$

$$\varepsilon_{y0} = \frac{\partial v_0}{\partial y}, \quad \kappa_y = -\frac{\partial^2 w}{\partial y^2} \qquad (2.68)$$

$$\gamma_{xy0} = \frac{\partial u_0}{\partial y} + \frac{\partial v_0}{\partial x}, \quad \kappa_{xy} = -\frac{\partial^2 w}{\partial x \partial y}$$

Thus, each strain component has a constant term, corresponding to the midsurface strain, and a term varying with z, corresponding to the change in curvature of the midsurface.

Furthermore, the substitution of strain components into the stress-strain relationship for plane stress produces

$$\sigma_x = \sigma_{x0}(x, y) + \frac{E}{1 - v^2} z(\kappa_x + v\kappa_y)$$

$$\sigma_y = \sigma_{y0}(x, y) + \frac{E}{1 - v^2} z(v\kappa_x + \kappa_y) \qquad (2.69)$$

$$\tau_{xy} = \tau_{xy0}(x, y) + 2Gz\kappa_{xy}$$

in which membrane stresses

$$\sigma_{x0} = \frac{E}{1 - v^2}(\varepsilon_{x0} + v\varepsilon_{y0})$$

$$\sigma_{y0} = \frac{E}{1 - v^2}(v\varepsilon_{x0} + \varepsilon_{y0}) \qquad (2.70)$$

$$\tau_{xy0} = G\gamma_{xy0}$$

The membrane stress resultants, such as in-plane forces and bending moments about the middle surface per unit width of section, are defined as (see Fig. 2.17)

$$(N_x, N_y, N_{xy}) = \int_{-\frac{h}{2}}^{\frac{h}{2}} (\sigma_x, \sigma_y, \tau_{xy}) dz = (h\sigma_{x0}, h\sigma_{y0}, h\tau_{xy0}) \qquad (2.71)$$

$$(M_x, M_y, M_{xy}) = \int_{-\frac{h}{2}}^{\frac{h}{2}} z(\sigma_x, \sigma_y, \tau_{xy}) dz$$

$$= (J_1(\kappa_x + \nu\kappa_y), J_1(\nu\kappa_x + \kappa_y), J_2\kappa_{xy})$$

$$= \left[-J_1\left(\frac{\partial^2 w}{\partial x^2} + \nu \frac{\partial^2 w}{\partial y^2} \right), -J_1\left(\nu \frac{\partial^2 w}{\partial x^2} + \frac{\partial^2 w}{\partial y^2} \right), -J_2 \frac{\partial^2 w}{\partial x \partial y} \right]$$

$$(2.72)$$

where

$$J_1 = \frac{Eh^3}{12(1-\nu^2)}, \quad J_2 = \frac{Gh^3}{6} = (1-\nu)J_1 \qquad (2.73)$$

It is evident that the shear forces are zero by virtue of the shear force resultants through the thickness. In practice, that is not reasonable. We can recover the shear forces from the equilibrium of thin plate.

Consider the microelement equilibrium, as illustrated in Fig. 2.18. In total, three force balances and two moment balances are involved:

$$\sum f_x = 0, \sum f_y = 0, \sum f_z = 0, \sum m_x = 0, \sum m_y = 0 \qquad (2.74)$$

from which we have

$$\frac{\partial N_x}{\partial x} + \frac{\partial N_{xy}}{\partial y} = 0$$

$$\frac{\partial N_{xy}}{\partial x} + \frac{\partial N_y}{\partial y} = 0 \qquad (2.75)$$

$$\frac{\partial Q_x}{\partial x} + \frac{\partial Q_y}{\partial y} + p = 0$$

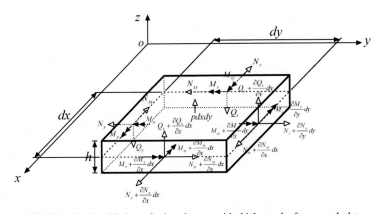

FIGURE 2.18 Equilibrium of microelement with thickness h of a general plate.

and

$$Q_x - \frac{\partial M_x}{\partial x} - \frac{\partial M_{xy}}{\partial y} = 0$$

$$Q_y - \frac{\partial M_{xy}}{\partial x} - \frac{\partial M_y}{\partial y} = 0$$

(2.76)

Substituting Eq. (2.76) into the third Eq. (2.75) yields

$$\frac{\partial^2 M_x}{\partial x^2} + 2\frac{\partial^2 M_{xy}}{\partial x \partial y} + \frac{\partial^2 M_y}{\partial y^2} + p(x, y) = 0$$

(2.77)

Subsequently, replacing the moment resultants in Eq. (2.77) with Eq. (2.72) gives

$$J_1 \nabla^4 w(x, y) = p$$

(2.78)

where

$$\nabla^4 = \frac{\partial^4}{\partial x^4} + 2\frac{\partial^4}{\partial x^2 \partial y^2} + \frac{\partial^4}{\partial y^4}$$

(2.79)

is the classic biharmonic operator.

As well, substituting Eq. (2.72) into Eq. (2.76) yields

$$Q_x = -J_1\left(\frac{\partial^3 w}{\partial x^3} + \frac{\partial^3 w}{\partial x \partial y^2}\right) = -J_1 \frac{\partial}{\partial x} \nabla^2 w$$

$$Q_y = -J_1\left(\frac{\partial^3 w}{\partial x^2 \partial y} + \frac{\partial^3 w}{\partial y^3}\right) = -J_1 \frac{\partial}{\partial y} \nabla^2 w$$

(2.80)

Generally, the displacements u_0 and v_0 are zero under pure bending conditions. Thus, Eq. (2.78) actually governs the lateral deflection of transversely loaded thin plate without axial deformations.

2.6.3 Boundary conditions

An exact solution of the governing plate equation must simultaneously satisfy the differential equation and the boundary conditions of any given plate problem. In the bending theory of thin plates, three internal force components are to be considered: bending moment, torsional moment, and transverse shear force. Similarly, the displacement components to be used in formulating the boundary conditions are lateral deflections and slope. Because the governing plate equation is a fourth-order differential equation, only two boundary conditions are required at each boundary point. By introducing the effective shear forces defined as

$$V_x = Q_x + \frac{\partial M_{xy}}{\partial y}$$

$$V_y = Q_y + \frac{\partial M_{xy}}{\partial x}$$

(2.81)

Kirchhoff thin plate theory reduces the number of internal forces to be considered from three to two. It should be mentioned that, at each corner of the rectangular plates, the aforementioned actions of torsional moments add up instead of canceling each other out, producing an additional corner force, as indicated in Fig. 2.19

$$R_o = 2M_{xy} \tag{2.82}$$

As a result, substituting Eqs. (2.72) and (2.80) into Eq. (2.81) gives

$$V_x = -J_1 \frac{\partial^3 w}{\partial x^3} - J_1(2 - \nu) \frac{\partial^3 w}{\partial x \partial y^2}$$

$$V_y = -J_1(2 - \nu) \frac{\partial^3 w}{\partial x^2 \partial y} - J_1 \frac{\partial^3 w}{\partial y^3} \tag{2.83}$$

For the case of a smooth boundary, it is assumed that **n** and **t** are the outward unit normal and tangential vectors to the boundary, respectively. It is clear from Fig. 2.20 that the vector **s** is related to the vector **n** in the form

$$\mathbf{n} = [n_1, n_2]^{\mathrm{T}}, \ \mathbf{s} = [t_1, t_2]^{\mathrm{T}} = [-n_2, n_1]^{\mathrm{T}} \tag{2.84}$$

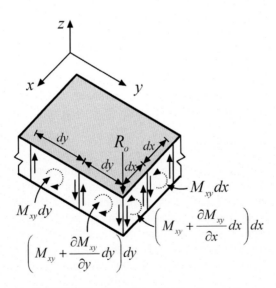

FIGURE 2.19   Edge effect of torsional moments.

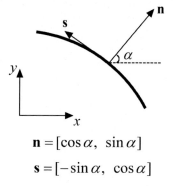

$$\mathbf{n} = [\cos\alpha, \ \sin\alpha]$$
$$\mathbf{s} = [-\sin\alpha, \ \cos\alpha]$$

FIGURE 2.20 Relation of normal and tangential vectors.

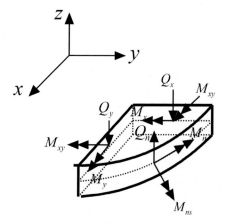

FIGURE 2.21 Forces on the curved boundary.

Subsequently, by considering the balance of the boundary microelement given in Fig. 2.21, the normal quantities on the boundary can be written as

$$\theta_n = w_{,i}n_i$$
$$M_n = -D\left[\nu w_{,ii} + (1-\nu)w_{,ij}n_in_j\right]$$
$$M_{nt} = -D(1-\nu)w_{,ij}n_it_j \tag{2.85}$$
$$Q_n = -Dw_{,ijj}n_i$$
$$V_n = Q_n + \frac{\partial M_{nt}}{\partial s} = -D\left[w_{,ijj}n_i + (1-\nu)w_{,ijk}n_it_jt_k\right]$$

TABLE 2.1 Typical boundary conditions for thin plate bending.

Type of support	Mathematical expressions
Simple support	$w = 0$, $M_n = 0$
Fixed edge	$w = 0$, $\theta_n = 0$
Free edge	$V_n = 0$, $M_n = 0$

where $\theta_n = \partial w/\partial n$ represents the normal rotation. M_n is the bending moment perpendicular to the normal direction, Q_n is the shear force perpendicular to the normal direction, M_{nt} is the torsional moment, and $V_n = Q_n + \partial M_{ns}/\partial s$ is the effect shear force perpendicular to the normal direction. As well, the comma denotes the partial differential to the spatial variable, i.e.,

$$w_{,ij} = \frac{\partial^2 w}{\partial x_i \partial x_j}, w_{,ijk} = \frac{\partial^2 w}{\partial x_i \partial x_j \partial x_k} \quad (i, j, k = 1, 2) \tag{2.86}$$

with $(x, y) = (x_1, x_2)$, and the repeated subscript represents the summation notation, i.e.,

$$w_{,i} n_i = w_{,1} n_1 + w_{,2} n_2 \tag{2.87}$$

Therefore, the boundary conditions on the smooth boundary of a thin plate can be described by two generalized displacement components, w and θ_n, and two generalized internal forces, V_n and M_n. Correspondingly, the mathematical expressions of boundary conditions are given in Table 2.1 for several support types.

2.7 Equations for piezoelectricity

Unlike conventional elastic materials, the piezoelectric material discovered by Pierre and Jacques Curie in 1880 [11] is a smart material, which can produce deformation when an electric potential difference is applied across it (indirect effect) and can generate an electric charge when it is stressed mechanically by a force (direct effect). Therefore, piezoelectric material is capable of acting as either a sensor or a transducer or both. Fig. 2.22 shows the direct and indirect piezoelectric effects in piezoelectric material.

Compared to classic linear elasticity that involves mechanical deformation only, piezoelectricity is a cross-coupling between elastic variables such as stress and strain and dielectric variables such as electric charge density and electric field [12−18]. Such coupling in piezoelectric materials is generally represented in electro-elastic constitutive equations. Hence, the basic governing equations of piezoelectricity [5,19−23] are more complex than those of classic linear elastic theory.

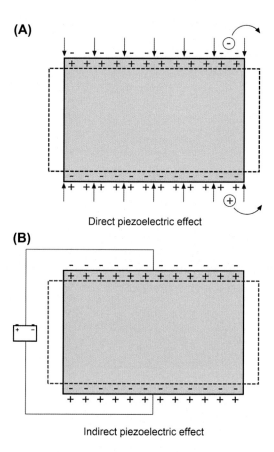

(A)

Direct piezoelectric effect

(B)

Indirect piezoelectric effect

FIGURE 2.22   Illustration of piezoelectric effects.

2.7.1 Governing equations

For a three-dimensional linear piezoelectric material under small field conditions, describing in the Cartesian coordinates (x, y, z) or (x_1, x_2, x_3), the equilibrium equations describing the relationship between the stress and the external volume force and the Gauss's law of electrostatics relating the distribution of electric charge to the resulting electric field can be written separately as

$$\frac{\partial \sigma_x}{\partial x} + \frac{\partial \tau_{xy}}{\partial y} + \frac{\partial \tau_{xz}}{\partial z} + b_x = 0$$

$$\frac{\partial \tau_{xy}}{\partial x} + \frac{\partial \sigma_y}{\partial y} + \frac{\partial \tau_{yz}}{\partial z} + b_y = 0$$

$$\frac{\partial \tau_{xz}}{\partial x} + \frac{\partial \tau_{yz}}{\partial y} + \frac{\partial \sigma_z}{\partial z} + b_z = 0$$

(2.88)

and

$$\frac{\partial D_x}{\partial x} + \frac{\partial D_y}{\partial y} + \frac{\partial D_z}{\partial z} = q \tag{2.89}$$

where σ_x, σ_y, σ_z, τ_{yz}, τ_{xz}, and τ_{xy} denote the stress components (N/m^2), b_x, b_y, and b_z are the body force components (N/m^3), D_x, D_y, and D_z are the electric displacement components (Coulomb/m^2), and q is the free electric volume charge (Coulomb/m^3).

Eqs. (2.88) and (2.89) can be rewritten in matrix notation as follows:

$$\mathbf{L}^T\boldsymbol{\sigma} + \mathbf{b} = 0$$
$$\nabla^T\mathbf{D} = q \tag{2.90}$$

where

$$\boldsymbol{\sigma} = \{\, \sigma_x \quad \sigma_y \quad \sigma_z \quad \tau_{yz} \quad \tau_{xz} \quad \tau_{xy} \,\}^T$$
$$\mathbf{b} = \{\, b_x \quad b_y \quad b_z \,\}^T \tag{2.91}$$
$$\mathbf{D} = \{\, D_x \quad D_y \quad D_z \,\}^T$$

and

$$\mathbf{L} = \begin{bmatrix} \dfrac{\partial}{\partial x} & 0 & 0 \\[2mm] 0 & \dfrac{\partial}{\partial y} & 0 \\[2mm] 0 & 0 & \dfrac{\partial}{\partial z} \\[2mm] 0 & \dfrac{\partial}{\partial z} & \dfrac{\partial}{\partial y} \\[2mm] \dfrac{\partial}{\partial z} & 0 & \dfrac{\partial}{\partial x} \\[2mm] \dfrac{\partial}{\partial y} & \dfrac{\partial}{\partial x} & 0 \end{bmatrix}, \quad \nabla = \begin{bmatrix} \dfrac{\partial}{\partial x} \\[2mm] \dfrac{\partial}{\partial y} \\[2mm] \dfrac{\partial}{\partial z} \end{bmatrix} \tag{2.92}$$

Subsequently, the kinematic relations define the strain tensor as a symmetric part of the gradient of mechanical displacements, whereas Maxwell's law states that the electric vector field is the negative gradient of electric potential, that is,

$$\varepsilon_x = \frac{\partial u}{\partial x}, \varepsilon_y = \frac{\partial u}{\partial y}, \varepsilon_z = \frac{\partial w}{\partial z},$$

$$\gamma_{yz} = \frac{\partial v}{\partial z} + \frac{\partial w}{\partial y}, \gamma_{xz} = \frac{\partial u}{\partial z} + \frac{\partial w}{\partial x}, \gamma_{xy} = \frac{\partial u}{\partial y} + \frac{\partial v}{\partial x} \tag{2.93}$$

and

$$E_x = -\frac{\partial \varphi}{\partial x}, \ E_y = -\frac{\partial \varphi}{\partial y}, E_z = -\frac{\partial \varphi}{\partial z} \qquad (2.94)$$

where ε_x, ε_y, ε_z, γ_{yz}, γ_{xz}, and γ_{xy} are the strain components (m/m); E_x, E_y, and E_z are the electric field components (Volt/m); u, v, and w are the elastic displacement components (m), and φ is the electric potential (Volt).

Correspondingly, Eqs. (2.93) and (2.94) can be displayed in matrix form as

$$\boldsymbol{\varepsilon} = \mathbf{Lu}$$
$$\mathbf{E} = -\nabla \varphi \qquad (2.95)$$

with

$$\boldsymbol{\varepsilon} = \left\{ \varepsilon_x \ \ \varepsilon_y \ \ \varepsilon_z \ \ \gamma_{yz} \ \ \gamma_{xz} \ \ \gamma_{xy} \right\}^{\mathrm{T}}$$
$$\mathbf{u} = \left\{ u \ \ v \ \ w \right\}^{\mathrm{T}} \qquad (2.96)$$
$$\mathbf{E} = \left\{ E_x \ \ E_y \ \ E_z \right\}^{\mathrm{T}}$$

Besides, the linear piezoelectric constitutive relations identifying the coupling of mechanical deformation and electric field (see Fig. 2.23) can be written in stress-charge form as

$$\sigma_k = c_{km}\varepsilon_m - e_{kj}E_j$$
$$D_i = e_{im}\varepsilon_m + \lambda^\varepsilon_{ij}E_j \qquad (2.97)$$

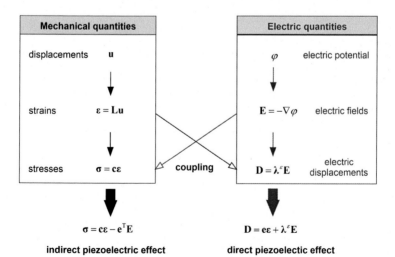

FIGURE 2.23 Illustration of coupling between mechanical and electrical quantities by piezoelectric effects.

or

$$\begin{bmatrix} \boldsymbol{\sigma} \\ \mathbf{D} \end{bmatrix} = \begin{bmatrix} \mathbf{c} & -\mathbf{e}^T \\ \mathbf{e} & \boldsymbol{\lambda}^\varepsilon \end{bmatrix} \begin{bmatrix} \boldsymbol{\varepsilon} \\ \mathbf{E} \end{bmatrix} \tag{2.98}$$

in which the piezoelectric constants are the elastic stiffness matrix $\mathbf{c}$ of size (6×6) (N/m^2) measured at zero or constant electric field, the piezoelectric coefficient matrix $\mathbf{e}$ of size (3×6) (Coulomb/m^2), and the dielectric permittivity matrix $\boldsymbol{\lambda}^\varepsilon$ of size (3×3) (Farad/m) measured at zero or constant strain. Moreover, the rule of change of subscripts (Kelvin-Voigt notation)

$$x(11) \to 1, y(22) \to 2, z(33) \to 3, yz(23) \to 4, xz(13) \to 5, xy(12) \to 6$$

is adopted here to convert the constitutive equations written in tensor form in Ref. [19] into Eq. (2.97) in the matrix notation. For example, using this rule the stress and strain vectors in Eqs. (2.91) and (2.96) correspond to

$$\boldsymbol{\sigma} = \{ \sigma_1 \quad \sigma_2 \quad \sigma_3 \quad \sigma_4 \quad \sigma_5 \quad \sigma_6 \}^T$$
$$\boldsymbol{\varepsilon} = \{ \varepsilon_1 \quad \varepsilon_2 \quad \varepsilon_3 \quad \varepsilon_4 \quad \varepsilon_5 \quad \varepsilon_6 \}^T$$

For orthotropic piezoelectric materials which have the poling direction along the 3-axis and the 1-axis and 2-axis are in the orthotropic plane of the material, there are 17 independent material constants, and the matrices of piezoelectric, stiffness, and dielectric constants are respectively given by

$$\mathbf{e} = \begin{bmatrix} 0 & 0 & 0 & 0 & e_{15} & 0 \\ 0 & 0 & 0 & e_{24} & 0 & 0 \\ e_{31} & e_{32} & e_{33} & 0 & 0 & 0 \end{bmatrix} \tag{2.99}$$

$$\mathbf{c} = \begin{bmatrix} c_{11} & c_{12} & c_{13} & 0 & 0 & 0 \\ c_{12} & c_{22} & c_{23} & 0 & 0 & 0 \\ c_{13} & c_{23} & c_{33} & 0 & 0 & 0 \\ 0 & 0 & 0 & c_{44} & 0 & 0 \\ 0 & 0 & 0 & 0 & c_{55} & 0 \\ 0 & 0 & 0 & 0 & 0 & c_{66} \end{bmatrix} \tag{2.100}$$

$$\boldsymbol{\lambda}^\varepsilon = \begin{bmatrix} \lambda_{11}^\varepsilon & 0 & 0 \\ 0 & \lambda_{22}^\varepsilon & 0 \\ 0 & 0 & \lambda_{33}^\varepsilon \end{bmatrix} \tag{2.101}$$

Typically, for many piezoelectric materials such as (lead zirconate titanate) PZT-4 ceramic, they can be treated as transversely isotropic. If the poling direction is denoted as the 3-axis and the 1-axis and 2-axis are in the isotropic plane of material, as indicated in Fig. 2.24, then there are 10 independent

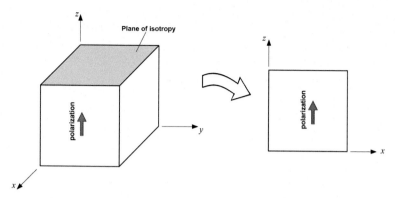

FIGURE 2.24 Illustration of a transverse isotropic linear piezoelectric material and the corresponding 2D modeling.

material constants. By setting $e_{24} = e_{15}$ and $e_{32} = e_{31}$ in Eq. (2.99), the reduced piezoelectric matrix can be expressed as

$$\mathbf{e} = \begin{bmatrix} 0 & 0 & 0 & 0 & e_{15} & 0 \\ 0 & 0 & 0 & e_{15} & 0 & 0 \\ e_{31} & e_{31} & e_{33} & 0 & 0 & 0 \end{bmatrix} \tag{2.102}$$

Correspondingly, we have the reduced stiffness matrix

$$\mathbf{c} = \begin{bmatrix} c_{11} & c_{12} & c_{13} & 0 & 0 & 0 \\ c_{12} & c_{11} & c_{13} & 0 & 0 & 0 \\ c_{13} & c_{13} & c_{33} & 0 & 0 & 0 \\ 0 & 0 & 0 & c_{44} & 0 & 0 \\ 0 & 0 & 0 & 0 & c_{44} & 0 \\ 0 & 0 & 0 & 0 & 0 & c_{66} \end{bmatrix} \tag{2.103}$$

by setting in Eq. (2.100)

$$c_{22} = c_{11}, c_{23} = c_{13}, c_{55} = c_{44}, c_{66} = \frac{c_{11} - c_{12}}{2} \tag{2.104}$$

Simultaneously, the reduced dielectric permittivity matrix can be obtained by letting $\lambda_{11}^{\varepsilon} = \lambda_{22}^{\varepsilon}$ in Eq. (2.101):

$$\boldsymbol{\lambda}^{\varepsilon} = \begin{bmatrix} \lambda_{11}^{\varepsilon} & 0 & 0 \\ 0 & \lambda_{11}^{\varepsilon} & 0 \\ 0 & 0 & \lambda_{33}^{\varepsilon} \end{bmatrix} \tag{2.105}$$

Specially, under the assumptions that the length along the y-direction is very small (plane stress) or the length along the y-direction is very large (plane strain), then the $x - z$ plane can be taken out as the reduced 2D piezoelectric model for the study of plane electromechanical phenomena [24]. For example, for plane strain condition, the reduced governing equations in the x-z plane can be derived by setting $\varepsilon_y = 0$, $\gamma_{yz} = 0$, $\gamma_{xy} = 0$, $E_y = 0$ as follows:

1. The equilibrium equations and Gauss's law of electrostatics

$$\begin{aligned} \mathbf{L}^T\boldsymbol{\sigma} + \mathbf{b} &= 0 \\ \nabla^T\mathbf{D} &= q \end{aligned} \tag{2.106}$$

where

$$\begin{aligned} \boldsymbol{\sigma} &= \{\,\sigma_x \quad \sigma_z \quad \tau_{xz}\,\}^T \\ \mathbf{b} &= \{\,b_x \quad b_z\,\}^T \\ \mathbf{D} &= \{\,D_x \quad D_z\,\}^T \end{aligned} \tag{2.107}$$

and

$$\mathbf{L} = \begin{bmatrix} \dfrac{\partial}{\partial x} & 0 \\[2mm] 0 & \dfrac{\partial}{\partial z} \\[2mm] \dfrac{\partial}{\partial z} & \dfrac{\partial}{\partial x} \end{bmatrix}, \quad \nabla = \begin{bmatrix} \dfrac{\partial}{\partial x} \\[2mm] \dfrac{\partial}{\partial z} \end{bmatrix} \tag{2.108}$$

2. The kinematic relations and Maxwell's law

$$\begin{aligned} \boldsymbol{\varepsilon} &= \mathbf{L}\mathbf{u} \\ \mathbf{E} &= -\nabla\varphi \end{aligned} \tag{2.109}$$

with

$$\begin{aligned} \boldsymbol{\varepsilon} &= \{\,\varepsilon_x \quad \varepsilon_z \quad \gamma_{xz}\,\}^T \\ \mathbf{u} &= \{\,u \quad w\,\}^T \\ \mathbf{E} &= \{\,E_x \quad E_z\,\}^T \end{aligned} \tag{2.110}$$

3. The constitutive relations with piezoelectric coupling

$$\begin{aligned} \boldsymbol{\sigma} &= \mathbf{c}\boldsymbol{\varepsilon} - \mathbf{e}^T\mathbf{E} \\ \mathbf{D} &= \mathbf{e}\boldsymbol{\varepsilon} + \boldsymbol{\lambda}^\varepsilon\mathbf{E} \end{aligned} \tag{2.111}$$

with

$$
\mathbf{c} = \begin{bmatrix} c_{11} & c_{13} & 0 \\ c_{13} & c_{33} & 0 \\ 0 & 0 & c_{44} \end{bmatrix}
$$

$$
\mathbf{e} = \begin{bmatrix} 0 & 0 & e_{15} \\ e_{31} & e_{33} & 0 \end{bmatrix} \qquad (2.112)
$$

$$
\boldsymbol{\lambda}^\varepsilon = \begin{bmatrix} \lambda_{11}^\varepsilon & 0 \\ 0 & \lambda_{33}^\varepsilon \end{bmatrix}
$$

On the other hand, for the plane stress state ($\sigma_y = 0$, $\tau_{yz} = 0$, $\tau_{xy} = 0$, $D_y = 0$), the electro-elastic constitutive equations can be obtained by several simple replacement operations in the material matrix Eq. (2.112), i.e., replacing c_{11} in Eq. (2.112) with $c_{11} - c_{12}^2/c_{11}$, c_{13} with $c_{13} - c_{12}c_{13}/c_{11}$, etc. Finally, the material matrix for plane stress can be written as

$$
\mathbf{c} = \begin{bmatrix} c_{11} - \dfrac{c_{12}^2}{c_{11}} & c_{13} - \dfrac{c_{12}c_{13}}{c_{11}} & 0 \\ c_{13} - \dfrac{c_{12}c_{13}}{c_{11}} & c_{33} - \dfrac{c_{13}^2}{c_{11}} & 0 \\ 0 & 0 & c_{44} \end{bmatrix}
$$

$$
\mathbf{e} = \begin{bmatrix} 0 & 0 & e_{15} \\ e_{31} - \dfrac{c_{12}e_{31}}{c_{11}} & e_{33} - \dfrac{c_{13}e_{31}}{c_{11}} & 0 \end{bmatrix} \qquad (2.113)
$$

$$
\boldsymbol{\lambda}^\varepsilon = \begin{bmatrix} \lambda_{11}^\varepsilon & 0 \\ 0 & \lambda_{33}^\varepsilon + \dfrac{e_{31}^2}{c_{11}} \end{bmatrix}
$$

Finally, substituting Eq. (2.95) into the constitutive relations (2.98), and then into the equilibrium equations and Gauss' law (2.90), we have the following set of partial differential equations in terms of the primary dependent variables $\mathbf{u}$ and φ:

$$
\mathbf{L}^T \mathbf{c} \mathbf{L} \mathbf{u} + \mathbf{L}^T \mathbf{e}^T \nabla \varphi + \mathbf{b} = 0 \qquad (2.114)
$$

$$
\nabla^T \mathbf{e} \mathbf{L} \mathbf{u} - \nabla^T \boldsymbol{\lambda}^\varepsilon \nabla \varphi = q \qquad (2.115)
$$

which can be further written in more compact form as

$$\begin{bmatrix} \mathbf{L}^T\mathbf{cL} & \mathbf{L}^T\mathbf{e}^T\nabla \\ \nabla^T\mathbf{eL} & -\nabla^T\boldsymbol{\lambda}^\varepsilon\nabla \end{bmatrix} \begin{Bmatrix} \mathbf{u} \\ \varphi \end{Bmatrix} + \begin{Bmatrix} \mathbf{b} \\ -q \end{Bmatrix} = 0 \qquad (2.116)$$

2.7.2 Boundary conditions

Boundary conditions in the electromechanical problem of piezoelectricity are "uncoupled." This means that the standard mechanical conditions are applied separately from the electrical conditions. It is assumed that the piezoelectric material occupying a domain Ω has a piecewise smoothed boundary Γ, which can be divided into two disjunctive parts dedicated to various essential and natural boundary conditions.

1. Mechanical boundary conditions

Mechanics conditions include the displacement boundary conditions given by

$$u = \overline{u}\,, w = \overline{w}, \quad \text{on } \Gamma_u \qquad (2.117)$$

and the surface traction boundary conditions

$$\begin{aligned} t_x &= \sigma_x n_x + \tau_{xz} n_z = \overline{t}_x \\ t_z &= \tau_{xz} n_x + \sigma_z n_z = \overline{t}_z \end{aligned}, \quad \text{on } \Gamma_t \qquad (2.118)$$

2. Electrical boundary conditions

Electrical conditions include the prescribed surface charge density boundary conditions given by

$$D_n = D_x n_x + D_z n_z = -\overline{\omega}, \quad \text{on } \Gamma_\omega \qquad (2.119)$$

and the prescribed electric potential boundary conditions given by

$$\varphi = \overline{\varphi}, \quad \text{on } \Gamma_\varphi \qquad (2.120)$$

where $\overline{u}, \overline{w}, \overline{\varphi}, \overline{t}_x, \overline{t}_z$, and $\overline{\omega}$ are specified values on the boundary; n_x and n_z are components of the unit outward normal vector, respectively. Note that $\Gamma_t \cup \Gamma_u = \Gamma_\omega \cup \Gamma_\varphi = \Gamma$ and $\Gamma_t \cap \Gamma_u = \Gamma_\omega \cap \Gamma_\varphi = \varnothing$.

2.8 Remarks

In this chapter, the fundamentals of mechanics for solids and structures are reviewed, as the related problems are frequently dealt with in this book. The review begins with classic 3D linear elasticity. Then, a solution in 3D space is reduced to a solution of 2D elasticity by introducing stress or stain assumptions that are true for two engineering problem types. Subsequently, the governing partial differential equations describing the Euler—Bernoulli beam and thin plate bending are derived from the theory of 3D linear elasticity by introducing specific displacement modes. Finally, the basic equations of 2D linear elasticity are extended to piezoelectric problems by including the piezoelectric effect. Readers with experience in mechanics may skip this chapter, but the chapter usefully introduces the terms used in the book.

References

[1]   A.J.M. Spencer, Continuum Mechanics, Longman, London, 1980.

[2]   S.C. Cowin, Continuum Mechanics of Anisotropic Materials, Springer, 2013.

[3]   M.H. Sadd, Elasticity: Theory, Applications, and Numerics, Academic Press, 2009.

[4]   S.P. Timoshenko, J.N. Goodier, Theory of Elasticity, Mcgraw Hill, 1970.

[5]   Q.H. Qin, Advanced Mechanics of Piezoelectricity, Higher Education Press and Springer, Beijing, 2013.

[6]   L.H. Donnell, Beams, Plates, and Shells, McGraw-Hill, New York, 1976.

[7]   P. Vannucci, Anisotropic Elasticity, Springer, 2018.

[8]   A.R. Damanpack, M. Bodaghi, H. Ghassemi, M. Sayehbani, Boundary element method applied to the bending analysis of thin functionally graded plates, Latin American Journal of Solids and Structures 10 (2013) 549—570.

[9]   S. Timoshenko, S. Woinowsky-Krieger, Theory of Plates and Shells, McGraw-Hill, New York, 1970.

[10]   E. Ventsel, Thin Plates and Shells Theory: Analysis, and Applications, CRC Press, Boca Raton, 2001.

[11]   P. Curie, J. Curie, Dévelopment, par pression, de l'électricité polaire dans les cristaux hémièdres à faces inclinées, Comptes Rendus de l'Académie des Sciences 91 (1980) 294—295.

[12]   Q.H. Qin, Thermoelectroelastic Green's function for a piezoelectric plate containing an elliptic hole, Mechanics of Materials 30 (1998) 21—29.

[13]   Q.H. Qin, Y.W. Mai, Crack growth prediction of an inclined crack in a half-plane thermopiezoelectric solid, Theoretical and Applied Fracture Mechanics 26 (1997) 185—191.

[14]   Q.H. Qin, Y.W. Mai, A closed crack tip model for interface cracks in thermopiezoelectric materials, International Journal of Solids and Structures 36 (1999) 2463—2479.

[15]   Q.H. Qin, S.W. Yu, An arbitrarily-oriented plane crack terminating at the interface between dissimilar piezoelectric materials, International Journal of Solids and Structures 34 (1997) 581—590.

[16]   S.W. Yu, Q.H. Qin, Damage analysis of thermopiezoelectric properties: Part I — crack tip singularities, Theoretical and Applied Fracture Mechanics 25 (1996) 263—277.

[17]   A.Y.T. Leung, X. Xu, Q. Gu, C.T.O. Leung, J.J. Zheng, The boundary layer phenomena in two-dimensional transversely isotropic piezoelectric media by exact symplectic expansion, International Journal for Numerical Methods in Engineering 69 (2007) 2381—2408.

[18]   C.H. Xu, Z.H. Zhou, X.S. Xu, A.Y.T. Leung, Electroelastic singularities and intensity factors for an interface crack in piezoelectric—elastic biomaterials, Applied Mathematical Modelling 39 (2015) 2721—2739.

[19]   T. Ikeda, Fundamentals of Piezoelectricity, Oxford University Press, New York, 1996.

[20] Q.H. Qin, Fracture Mechanics of Piezoelectric Materials, WIT Press, Southampton, 2001.

[21] Q.H. Qin, Y.W. Mai, Thermoelectroelastic Green's function and its application for bimaterial of piezoelectric materials, Archive of Applied Mechanics 68 (1998) 433−444.

[22] Q.H. Qin, Y.W. Mai, S.W. Yu, Effective moduli for thermopiezoelectric materials with microcracks, International Journal of Fracture 91 (1998) 359−371.

[23] Q.H. Qin, Y.W. Mai, S.W. Yu, Some problems in plane thermopiezoelectric materials with holes, International Journal of Solids and Structures 36 (1999) 427−439.

[24] H.A. Sosa, M.A. Castro, Electroelastic analysis of piezoelectric laminated structures, Applied Mechanics Reviews 46 (1993) 21−28.

Chapter 3

Basics of fundamental solutions and radial basis functions

Chapter outline

3.1 Introduction

As we have discussed in Chapter 1, to find an approximate solution of a problem governed by partial differential equations and boundary conditions, the unknown field function is first approximated by trail functions; the related theoretical formulation can then be applied to build the final discrete system of linear equations. In this book, the fundamental solutions of a problem and the radial basis functions are treated as trail functions for the implementation of the present meshless numerical technique. This chapter provides their basic definitions before we discuss any of theoretical foundation and applications of the present meshless method.

The chapter starts by illustrating a generalized partial differential operator, and its fundamental solutions representing the response of a point source in an infinite domain are then described. Subsequently, the concept of radial basis functions and its interpolation application are introduced.

3.2 Basic concept of fundamental solutions

3.2.1 Partial differential operators

Denoting $\mathbb{R}^n$ the Euclidean space with dimension n, the k-th partial differential operator of a function $u(\mathbf{x})$ can be defined by [1].

Methods of Fundamental Solutions in Solid Mechanics. https://doi.org/10.1016/B978-0-12-818283-3.00003-8

$$D^{\mathbf{k}}[u(\mathbf{x})] = D^{(k_1,k_2,\ldots,k_n)}[u(\mathbf{x})] = \frac{\partial^{k_1+k_2+\cdots+k_n} u(\mathbf{x})}{\partial x_1^{k_1} \partial x_2^{k_2} \cdots \partial x_n^{k_n}}, \quad \mathbf{x} = (x_1, x_2, \ldots, x_n) \in \mathbb{R}^n$$

(3.1)

where $\mathbf{k} = (k_1, k_2, \ldots, k_n)$, $k_1, k_2, \ldots, k_n$ are nonnegative integers. Specially, if the integer k_i is zero, the partial derivative with variable x_i is omitted. For example, for a two-dimensional Euclidean space, $n = 2$, we have this:

$$D^{0,0}[u(\mathbf{x})] = u(\mathbf{x}), \quad \text{for } \mathbf{k} = (0,0)$$

(3.2)

$$D^{1,0}[u(\mathbf{x})] = \frac{\partial u(\mathbf{x})}{\partial x_1}, \quad \text{for } \mathbf{k} = (1,0)$$

$$D^{0,1}[u(\mathbf{x})] = \frac{\partial u(\mathbf{x})}{\partial x_2}, \quad \text{for } \mathbf{k} = (0,1)$$

(3.3)

$$D^{1,1}[u(\mathbf{x})] = \frac{\partial^2 u(\mathbf{x})}{\partial x_1 \partial x_2}, \quad \text{for } \mathbf{k} = (1,1)$$

$$D^{2,0}[u(\mathbf{x})] = \frac{\partial^2 u(\mathbf{x})}{\partial x_1^2}, \quad \text{for } \mathbf{k} = (2,0)$$

(3.4)

$$D^{0,2}[u(\mathbf{x})] = \frac{\partial^2 u(\mathbf{x})}{\partial x_2^2}, \quad \text{for } \mathbf{k} = (0,2)$$

Then, an arbitrary p-order linear differential operator $\mathbf{L}$ of function $u(\mathbf{x})$ in terms of the independent spatial variables $\mathbf{x} = (x_1, x_2, \ldots, x_n)$ can be expressed as follows [1]:

$$\mathbf{L}u(\mathbf{x}) \equiv \mathbf{L}\{D[u(\mathbf{x})]\} = \sum_{s \le p} a_{\mathbf{k}}(\mathbf{x}) D^{\mathbf{k}}[u(\mathbf{x})]$$

(3.5)

where the coefficients $a_{\mathbf{k}}(\mathbf{x}) = a_{(k_1,k_2,\ldots,k_n)}(\mathbf{x})$ are functions of the spatial variable $\mathbf{x}$, and $s = \sum_{i=1}^{n} k_i$. Giving an example, the most general second-order linear partial differential operator $\mathbf{L}$ in two-dimensional space can be written as this:

$$\mathbf{L}u(\mathbf{x}) = \sum_{s \le 2} a_{\mathbf{k}}(\mathbf{x}) D^{\mathbf{k}}[u(\mathbf{x})]$$

$$= \sum_{s=0} a_{\mathbf{k}}(\mathbf{x}) D^{\mathbf{k}}[u(\mathbf{x})] + \sum_{s=1} a_{\mathbf{k}}(\mathbf{x}) D^{\mathbf{k}}[u(\mathbf{x})] + \sum_{s=2} a_{\mathbf{k}}(\mathbf{x}) D^{\mathbf{k}}[u(\mathbf{x})]$$

$$= a_{0,0}(\mathbf{x}) D^{0,0}[u(\mathbf{x})] + a_{1,0}(\mathbf{x}) D^{1,0}[u(\mathbf{x})] + a_{0,1}(\mathbf{x}) D^{0,1}[u(\mathbf{x})]$$

$$+ a_{2,0}(\mathbf{x}) D^{2,0}[u(\mathbf{x})] + a_{1,1}(\mathbf{x}) D^{1,1}[u(\mathbf{x})] + a_{0,2}(\mathbf{x}) D^{0,2}[u(\mathbf{x})]$$

(3.6)

$$= a_{0,0}(\mathbf{x}) u(\mathbf{x}) + a_{1,0}(\mathbf{x}) \frac{\partial u(\mathbf{x})}{\partial x_1} + a_{0,1}(\mathbf{x}) \frac{\partial u(\mathbf{x})}{\partial x_2}$$

$$+ a_{2,0}(\mathbf{x}) \frac{\partial^2 u(\mathbf{x})}{\partial x_1^2} + a_{1,1}(\mathbf{x}) \frac{\partial^2 u(\mathbf{x})}{\partial x_1 \partial x_2} + a_{0,2}(\mathbf{x}) \frac{\partial^2 u(\mathbf{x})}{\partial x_2^2}$$

From the preceding, some common operators can be obtained as the special cases:

- Laplacian operator

$$\nabla^2 u(\mathbf{x}) = \frac{\partial^2 u(\mathbf{x})}{\partial x_1^2} + \frac{\partial^2 u(\mathbf{x})}{\partial x_2^2} = D^{(2,0)}[u(\mathbf{x})] + D^{(0,2)}[u(\mathbf{x})] \tag{3.7}$$

- Helmholtz operator

$$\begin{aligned} \nabla^2 u(\mathbf{x}) + \lambda^2 u(\mathbf{x}) &= \frac{\partial^2 u(\mathbf{x})}{\partial x_1^2} + \frac{\partial^2 u(\mathbf{x})}{\partial x_2^2} + \lambda^2 u(\mathbf{x}) \\ &= D^{(2,0)}[u(\mathbf{x})] + D^{(0,2)}[u(\mathbf{x})] + \lambda^2 u(\mathbf{x}) \end{aligned} \tag{3.8}$$

- Modified Helmholtz operator

$$\begin{aligned} \nabla^2 u(\mathbf{x}) - \lambda^2 u(\mathbf{x}) &= \frac{\partial^2 u(\mathbf{x})}{\partial x_1^2} + \frac{\partial^2 u(\mathbf{x})}{\partial x_2^2} - \lambda^2 u(\mathbf{x}) \\ &= D^{(2,0)}[u(\mathbf{x})] + D^{(0,2)}[u(\mathbf{x})] - \lambda^2 u(\mathbf{x}) \end{aligned} \tag{3.9}$$

- Convection-diffusion operator

$$\begin{aligned} &\nabla^2 u(\mathbf{x}) + b_1(\mathbf{x})\frac{\partial u(\mathbf{x})}{\partial x_1} + b_2(\mathbf{x})\frac{\partial u(\mathbf{x})}{\partial x_2} - c(\mathbf{x})u(\mathbf{x}) \\ &= \frac{\partial^2 u(\mathbf{x})}{\partial x_1^2} + \frac{\partial^2 u(\mathbf{x})}{\partial x_2^2} + b_1(\mathbf{x})\frac{\partial u(\mathbf{x})}{\partial x_1} + b_2(\mathbf{x})\frac{\partial u(\mathbf{x})}{\partial x_2} - c(\mathbf{x})u(\mathbf{x}) \\ &= D^{(2,0)}[u(\mathbf{x})] + D^{(0,2)}[u(\mathbf{x})] + b_1(\mathbf{x})D^{(1,0)}[u(\mathbf{x})] \\ &\quad + b_2(\mathbf{x})D^{(0,1)}[u(\mathbf{x})] - c(\mathbf{x})D^{(0,0)}u(\mathbf{x}) \end{aligned} \tag{3.10}$$

Besides, the adjoint operator $\mathbf{L}^*$ of $\mathbf{L}$ is formally defined in Ref. [1] as shown:

$$\mathbf{L}^* u(\mathbf{x}) = \sum_{s \leq p} (-1)^s D^{\mathbf{k}}[a_{\mathbf{k}}(\mathbf{x})u(\mathbf{x})] \tag{3.11}$$

Further, a linear operator is said to be self-adjoint only if $\mathbf{L} = \mathbf{L}^*$. For example, for the most common Laplacian operator ∇^2 in $\mathbb{R}^3$,

$$\nabla^2 = D^{(2,0,0)} + D^{(0,2,0)} + D^{(0,0,2)} = \frac{\partial^2}{\partial x_1^2} + \frac{\partial^2}{\partial x_2^2} + \frac{\partial^2}{\partial x_3^2} \quad \text{in } \mathbb{R}^3 \qquad (3.12)$$

This is obviously self-adjoint.

3.2.2 Fundamental solutions

From the mathematical point of view, a fundamental solution (or free space Green's function) for a linear partial differential operator $\mathbf{L}$ is defined by the solution of the following inhomogeneous differential equation [1]:

$$\mathbf{L}^* G^*(\mathbf{x}, \mathbf{x}_s) = \delta(\mathbf{x}, \mathbf{x}_s) \qquad (3.13)$$

where $\mathbf{L}^*$ is the adjoint differential operator of $\mathbf{L}$, $\mathbf{x} = (x_1, x_2,\ldots, x_n)$ represents field point, $\mathbf{x}_s = (x_{s1}, x_{s2,\ldots}, x_{sn}) \in \mathbb{R}^n$ denotes a given source point at which a unit concentrated source is applied, and $\delta(\mathbf{x}, \mathbf{x}_s)$ is the Dirac delta function, which is nonzero at the point $\mathbf{x}_s$ but zero elsewhere, as indicated in Fig. 3.1.

Since Dirac delta function is used extensively in the development of fundamental solutions, it is worth revising it here. Dirac delta function in a two-dimensional case on a coordinate plane at $\mathbf{x} = (x_1, x_2) \in \mathbb{R}^2$ is defined by

$$\int_{-\infty}^{\infty} \int_{-\infty}^{\infty} \delta(\mathbf{x}, \mathbf{x}_s) dx_1 dx_2 = 1 \qquad (3.14)$$

and

$$\delta(\mathbf{x}, \mathbf{x}_s) = 0 \quad \text{for } \mathbf{x} \neq \mathbf{x}_s \qquad (3.15)$$

from which the following shifting property can be obtained:

$$\int_{-\infty}^{\infty} \int_{-\infty}^{\infty} \delta(\mathbf{x}, \mathbf{x}_s) f(\mathbf{x}) dx_1 dx_2 = f(\mathbf{x}_s) \qquad (3.16)$$

where $f(\mathbf{x})$ is a two-dimensional integrable function.

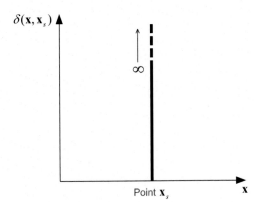

FIGURE 3.1   Dirac delta function.

Based on the concept of Dirac delta function, some important properties of the fundamental solution $G^*(\mathbf{x}, \mathbf{x}_s)$ for the linear operator $\mathbf{L}$ can be obtained as follows:

- $G^*(\mathbf{x}, \mathbf{x}_s)$ is defined everywhere, except at $\mathbf{x} = \mathbf{x}_s$, where it is singular.
- $\mathbf{L}^*G^*(\mathbf{x}, \mathbf{x}_s) = 0$ for the case at $\mathbf{x} \neq \mathbf{x}_s$.
- The reciprocity relation $G^*(\mathbf{x}, \mathbf{x}_s) = G^*(\mathbf{x}_s, \mathbf{x})$ holds.

For example, for the two-dimensional Laplacian operator ∇^2, which is linear and self-adjoint, its fundamental solution represents the static response at the field point $\mathbf{x} = (x_1, x_2)$ when a unit point source is applied at the source point $\mathbf{x}_s = (x_{s1}, x_{s2})$ in an infinite domain (Fig. 3.2) and satisfies

$$\nabla^2 G^*(\mathbf{x}, \mathbf{x}_s) = \delta(\mathbf{x}, \mathbf{x}_s) \tag{3.17}$$

By applying Fourier transform to Eq. (3.17), one has the following (see Ref. [1] for details):

$$G^*(\mathbf{x}, \mathbf{x}_s) = -\frac{1}{2\pi} \ln r \tag{3.18}$$

where $r = |\mathbf{x} - \mathbf{x}_s|$ represents the Euclidean distance from the field point $\mathbf{x} = (x_1, x_2)$ to the source point $\mathbf{x}_s = (x_{s1}, x_{s2})$, as indicated in Fig. 3.2.

Mathematically, the Cartesian components r_i of this distance can be written as

$$r_i = x_i - x_{si}, \quad i = 1, 2 \tag{3.19}$$

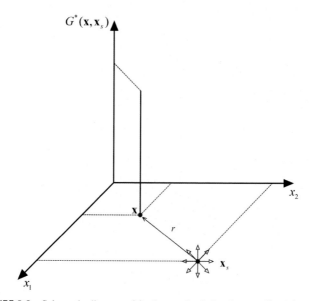

FIGURE 3.2 Schematic diagram of fundamental solution for two-dimensional case.

Using the summation convention to the repeated subscript, the distance r can be rewritten as

$$r = \sqrt{r_i r_i} = \sqrt{r_1^2 + r_2^2} \tag{3.20}$$

From the preceding, the spatial derivatives of r can be given by the chain rule of differentiation (higher-order derivations can be found in Appendix A):

$$\frac{\partial r}{\partial x_i} = \frac{r_i}{r} \tag{3.21}$$

$$\frac{\partial^2 r}{\partial x_i^2} = \frac{r^2 - r_i^2}{r^3} \tag{3.22}$$

To clearly illustrate the variation of fundamental solution Eq. (3.18) in the two-dimensional infinite plane, let us consider a special case where a unit point source is applied at the source point $\mathbf{x}_s = 0$. In such case, Eq. (3.18) reduces to this:

$$G^*(\mathbf{x}, 0) = -\frac{1}{2\pi} \ln \sqrt{x_1^2 + x_2^2} = -\frac{1}{4\pi} \ln(x_1^2 + x_2^2) \tag{3.23}$$

Fig. 3.3 displays the distribution of the special solution Eq. (3.23), and it is observed from Fig. 3.3 that the value of $G^*(\mathbf{x}, 0)$ decreases gradually with the increase of Euclidian distance r.

Specially, when the field point does not overlap with the source point, the fundamental solution analytically satisfies the following homogeneous partial differential equation:

$$\nabla^2 G^*(\mathbf{x}, \mathbf{x}_s) = 0, \quad \mathbf{x} \neq \mathbf{x}_s \tag{3.24}$$

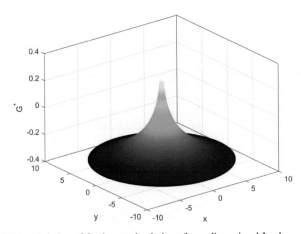

FIGURE 3.3 Variation of fundamental solution of two-dimensional Laplace equation.

This feature is the basis to build the methods of fundamental solution [2−7] and the fundamental-solution-based hybrid finite element methods [8−25], which use the linear combination of fundamental solution of problem to approximate desired field variable.

In addition, because the fundamental solution Eq. (3.18) is only dependent on the distance r between the source point and the field point, it satisfies the reciprocity relation:

$$G^*(\mathbf{x}, \mathbf{x}_s) = G^*(\mathbf{x}_s, \mathbf{x}) \tag{3.25}$$

3.3 Radial basis function interpolation

3.3.1 Radial basis functions

Definition 3.3.1. A function $\phi: \mathbb{R}^n \to \mathbb{R}$ is called radial function centered at the origin provided it is a univariate function in terms of variable r, so we have this:

$$\phi(\mathbf{x}) = \phi(r), \quad \mathbf{x} \in \mathbb{R}^n \tag{3.26}$$

where the real-valued variable

$$r = \|\mathbf{x}\| \in \mathbb{R} \tag{3.27}$$

is a norm on $\mathbb{R}^n$, usually an Euclidean norm. Obviously, Definition 3.3.1 implies that for a radial function ϕ, the relation

$$\phi(\|\mathbf{x}_1\|) = \phi(\|\mathbf{x}_2\|), \quad \mathbf{x}_1, \mathbf{x}_2 \in \mathbb{R}^n \tag{3.28}$$

holds only if

$$\|\mathbf{x}_1\| = \|\mathbf{x}_2\| \tag{3.29}$$

For example, a well-represented radial function, Gaussian radial function, is centered at the origin in $\mathbb{R}^2$ (two-dimensional space). It can be written as

$$\phi(\mathbf{x}) = \phi(r) = e^{-cr^2}, \quad \mathbf{x} = (x, y) \in \mathbb{R}^2, \ r \in \mathbb{R} \tag{3.30}$$

where $c > 0$ is the shape parameter, and $r = x^2 + y^2$.

Fig. 3.4 shows the graphs of two Gaussian radial functions, one with the shape parameter $c = 1$ and another with $c = 4$. It is found that a smaller value of c causes the surface to become flatter, while larger c leads to a sharper radial function and localizes its influence, which means the choice of c has a profound influence on the behavior of the Gaussian radial function. Additionally, the first- and second-order derivatives of the Gaussian radial function with $c = 1$ can be obtained using the chain rule of differentiation (see Appendix A):

$$\frac{\partial \phi}{\partial x} = \frac{d\phi}{dr} \frac{\partial r}{\partial x} = -2xe^{-r^2} \tag{3.31}$$

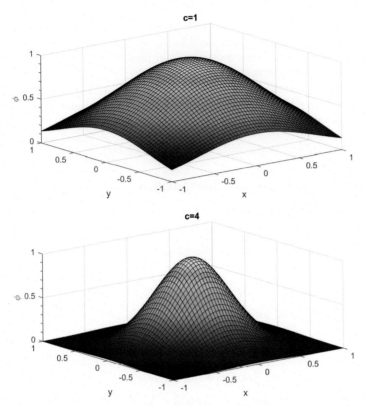

FIGURE 3.4   Variations of Gaussian radial functions with $c = 1$ and $c = 4$ centered at the origin.

$$\frac{\partial^2 \phi}{\partial x^2} = \frac{d^2 \phi}{dr^2}\left(\frac{\partial r}{\partial x}\right)^2 + \frac{d\phi}{dr}\frac{\partial^2 r}{\partial x^2} = -2e^{-r^2} + 4x^2 e^{-r^2} \qquad (3.32)$$

Their variations are evaluated and plotted in Figs. 3.5 and 3.6. From these two figures, one can find that the values of the first- and second-order derivatives $\partial \phi/\partial x$ and $\partial^2 \phi/\partial x^2$ are not constant but change very smoothly, suggesting the Gaussian radial function has higher-order continuity.

More generally, a smooth radial function can be defined, whose value depends only on the Euclidean norm of a vector from a central point $\widetilde{\mathbf{x}}$, called here a center point or reference point, to a field point $\mathbf{x}$ (i.e., Fig. 3.7 in $\mathbb{R}^2$), so

$$\phi(\mathbf{x}, \widetilde{\mathbf{x}}) = \phi(r) \qquad (3.33)$$

where

$$r = \|\mathbf{x} - \widetilde{\mathbf{x}}\| \qquad (3.34)$$

Thus, the radial function with simple shape can be easily applied in multidimensional space without causing significant extra effort. Besides, the evaluation of its derivatives is simpler than that for the traditional moving least

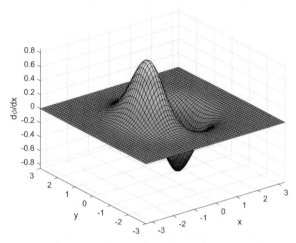

FIGURE 3.5 The first-order derivative of Gaussian radial function with $c = 1$ centered at the origin.

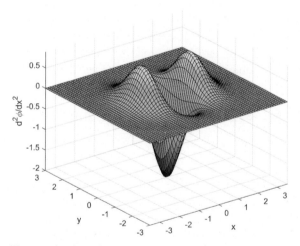

FIGURE 3.6 The second-order derivative of Gaussian radial function with $c = 1$ centered at the origin.

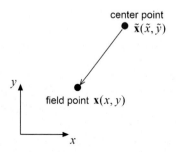

FIGURE 3.7 Definition of Euclidean distance between a center point and a field point.

TABLE 3.1 Some common radial functions.

Piecewise smooth RBFs	Power spline (PS)	r^{2n+1}
	Thin plate spline (TPS)	$r^{2n}\ln r$
Infinitely smooth RBFs	Multiquadric (MQ)	$\sqrt{r^2+c^2}$
	Gaussian (GS)	e^{-cr^2}

squares approximation [26,27]. Table 3.1 tabulates some common radial functions, in which $n = 1, 2, 3,\ldots$. To highlight the feature of radial functions, some typical radial functions listed in Table 3.1 are plotted in Fig. 3.8, from which it is observed that the value of these radial functions depends on the distance r only.

Since the radial functions are only Euclidean norm dependent, they have distinctive properties of being invariant under all Euclidean transformations (i.e., translations, rotations, and reflections) and being insensitive to the dimensions of the space. This brings the advantage on avoidance of multivariate functions in practice, whose complexity will increase with the increase of space dimensions. Moreover, a series of linear independent radial basis functions (RBFs) can be easily generated by radial function with different center points. Such ability of generating a large number of linearly independent basis functions is superior to that of polynomial functions, which makes the application of RBFs to the solution of the scattered data interpolation problem straightforward and beneficial. So far, RBFs have been used for

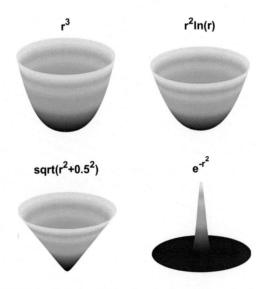

FIGURE 3.8 Typical radial functions in the two-dimensional space.

scattered data fitting and multidimensional data interpolation and were later applied to analyze the partial differential equations [28].

3.3.2 Radial basis function interpolation

As we have discussed in Section 3.3.1, a key feature of RBFs is its convenience to approximate the target function. The only geometric parameter that is used in RBFs approximation is the pairwise distance between points. Since distances are easy to obtain at any space dimensions, higher dimensions do not add more difficulties for computation. In practice, the RBFs approximation works with points scattered throughout the domain of interest, and a function can be interpolated by the linear combination of RBFs centered at series of points $\widetilde{\mathbf{x}}_i$ $(i = 1, ..., n)$:

$$f(\mathbf{x}) \approx \sum_{i=1}^{N} \alpha_i \phi(\mathbf{x}, \widetilde{\mathbf{x}}_i) \tag{3.35}$$

where f is the target function to be approximated, α_i is the interpolating coefficients, and N is the number of interpolating center points $\widetilde{\mathbf{x}}_i$.

In matrix form, Eq. (3.35) can be rewritten as shown:

$$\begin{bmatrix} \phi(\mathbf{x}_1, \widetilde{\mathbf{x}}_1) & \phi(\mathbf{x}_1, \widetilde{\mathbf{x}}_2) & \cdots & \phi(\mathbf{x}_1, \widetilde{\mathbf{x}}_N) \\ \phi(\mathbf{x}_2, \widetilde{\mathbf{x}}_1) & \phi(\mathbf{x}_2, \widetilde{\mathbf{x}}_2) & \cdots & \phi(\mathbf{x}_2, \widetilde{\mathbf{x}}_N) \\ \vdots & \vdots & \ddots & \vdots \\ \phi(\mathbf{x}_N, \widetilde{\mathbf{x}}_1) & \phi(\mathbf{x}_N, \widetilde{\mathbf{x}}_2) & \cdots & \phi(\mathbf{x}_N, \widetilde{\mathbf{x}}_N) \end{bmatrix} \begin{bmatrix} \alpha_1 \\ \alpha_2 \\ \vdots \\ \alpha_N \end{bmatrix} = \begin{bmatrix} f(\mathbf{x}_1) \\ f(\mathbf{x}_2) \\ \vdots \\ f(\mathbf{x}_N) \end{bmatrix} \tag{3.36}$$

or

$$\mathbf{\Phi}\boldsymbol{\alpha} = \mathbf{f} \tag{3.37}$$

From the preceding, one can determine all unknowns by solving this linear system of equations if the target function is given:

$$\boldsymbol{\alpha} = \mathbf{\Phi}^{-1}\mathbf{f} \tag{3.38}$$

To illustrate the accuracy and stability of RBFs approximation, we consider the following two-dimensional test function:

$$\begin{aligned} f(x, y) = {} & \frac{-751\pi^2}{144} \sin\frac{\pi x}{6} \sin\frac{7\pi x}{4} \sin\frac{3\pi y}{4} \sin\frac{5\pi y}{4} \\ & + \frac{7\pi^2}{12} \cos\frac{\pi x}{6} \cos\frac{7\pi x}{4} \sin\frac{3\pi y}{4} \sin\frac{5\pi y}{4} \\ & + \frac{15\pi^2}{8} \sin\frac{\pi x}{6} \sin\frac{7\pi x}{4} \cos\frac{3\pi y}{4} \cos\frac{5\pi y}{4} \end{aligned} \tag{3.39}$$

whose variation in a unit square domain is shown in Fig. 3.9.

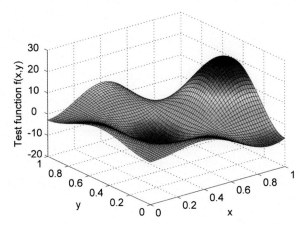

FIGURE 3.9 Variation of the test function in the unit square domain.

Additionally, the average relative error of the test function f (Arerr(f)) defined in Eq. (1.51) is employed to investigate the accuracy of the RBF approximation.

In the computation, the test function $f(\mathbf{x})$ is applied in a unit square domain and is approximated by seven RBFs listed in Table 3.2. The convergence of RBFs approximation for the given smooth test function is investigated with different numbers of regularly distributed interpolation points. Final numerical results of average relative error with $101 \times 101 = 10,201$ computing points regularly distributed in the domain and related condition numbers of the solving matrix $\mathbf{\Phi}$ are shown in Figs. 3.10 and 3.11, from which we can see that the power spline (PS) and thin plate spline (TPS) basis functions have slower convergence rate, compared to the multiquadric (MQ) basis function. Moreover, the shape parameter c in the MQ basis function may largely affect the stability and convergence of numerical results. And the GS basis

TABLE 3.2 List of radial basis functions used for approximation.

RBF1	r
RBF2	r^3
RBF3	$r^2 \ln r$
RBF4	$r^4 \ln r$
RBF5	$\sqrt{r^2 + c^2}$ with $c = 0.5$
RBF6	$\sqrt{r^2 + c^2}$ with $c = 0.5$
RBF7	e^{-r^2}

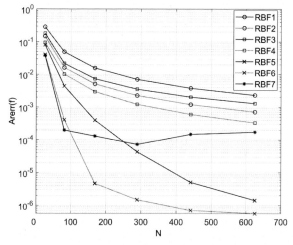

FIGURE 3.10 Variation of average relative error in terms of the number of interpolating points.

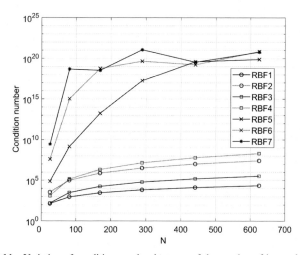

FIGURE 3.11 Variation of condition number in terms of the number of interpolating points.

function shows the worst results in all RBFs used here. The variations of condition number also show that the PS and TPS basis functions and the MQ basis function with small shape parameter have good stability, as the number of interpolation points increase. In addition, the results in Fig. 3.10 also show that the higher-order PS and TPS basis functions can bring better accuracy than the lower-order ones, while the related condition number becomes larger. Meanwhile, it is necessary to point out that the TPS basis function shows little improvement over the PS basis function.

Furthermore, as pointed out in Ref. [29], the MQ basis function converges exponentially, while the TPS and PS basis functions converge in $O(h|logh|)$ and $O(h^{1/2})$, respectively, in the data spacing $\{\widetilde{\mathbf{x}}_i\} \in \mathbb{R}^n$ and $h = \max\limits_{\mathbf{x} \in \mathbb{R}^n} \min\limits_{\widetilde{\mathbf{x}} \in \{\widetilde{\mathbf{x}}_i\}} \|\widetilde{\mathbf{x}} - \mathbf{x}\|$. However, when applying the MQ basis function, the shape parameter needs to be chosen carefully because the accuracy of MQ interpolants can vary by three orders of magnitude over a small range (typically $0 < c \le 10$).

3.4 Remarks

This chapter reviews some basic concepts of the fundamental solutions and the RBFs, aiming at laying the groundwork for the development of the present meshless method. The method is established by coupling the classic method of fundamental solutions and the radial basis function interpolation for inhomogeneous boundary value problems, including the beam bending problem, the thin plate bending problem, two-dimensional thermoelasticity, functionally graded elasticity, etc., which are described in the next few chapters.

References

[1] P.K. Kythe, Fundamental Solutions for Differential Operators and Applications, Birkhauser, Boston, 1996.

[2] C.S. Chen, A. Karageorghis, Y.S. Smyrlis, The Method of Fundamental Solutions: A Meshless Method, Dynamic Publishers, 2008.

[3] G. Fairweather, A. Karageorghis, The method of fundamental solutions for elliptic boundary value problems, Advances in Computational Mathematics 9 (1998) 69−95.

[4] P.A. Ramachandran, Method of fundamental solutions: singular value decomposition analysis, Communications in Numerical Methods in Engineering 18 (2002) 789−801.

[5] M.A. Golberg, C.S. Chen, The Method of Fundamental Solution for Potential, Helmholtz and Diffusion Problems, Computational Mechanics Publications, Southampton, 1998.

[6] L. Marin, D. Lesnic, The method of fundamental solutions for the Cauchy problem in two-dimensional linear elasticity, International Journal of Solids and Structures 41 (2004) 3425−3438.

[7] L. Marin, D. Lesnic, The method of fundamental solutions for nonlinear functionally graded materials, International Journal of Solids and Structures 44 (2007) 6878−6890.

[8] H. Wang, L.L. Cao, Q.H. Qin, Hybrid graded element model for nonlinear functionally graded materials, Mechanics of Advanced Materials and Structures 19 (2012) 590−602.

[9] H. Wang, Y.T. Gao, Q.H. Qin, Green's function based finite element formulations for isotropic seepage analysis with free surface, Latin American Journal of Solids and Structures 12 (2015) 1991−2005.

[10] H. Wang, M.Y. Han, F. Yuan, Z.R. Xiao, Fundamental-solution-based hybrid element model for nonlinear heat conduction problems with temperature-dependent material properties, Mathematical Problems in Engineering 2013 (2013) 8, 695457.

[11] H. Wang, Q.H. Qin, Hybrid FEM with fundamental solutions as trial functions for heat conduction simulation, Acta Mechanica Solida Sinica 22 (2009) 487−498.

[12] H. Wang, Q.H. Qin, Fundamental-solution-based finite element model for plane orthotropic elastic bodies, European Journal of Mechanics − A: Solids 29 (2010) 801−809.

[13] H. Wang, Q.H. Qin, FE approach with Green's function as internal trial function for simulating bioheat transfer in the human eye, Archives of Mechanics 62 (2010) 493−510.

[14] H. Wang, Q.H. Qin, Special fiber elements for thermal analysis of fiber-reinforced composites, Engineering Computations 28 (2011) 1079−1097.

[15] H. Wang, Q.H. Qin, Fundamental-solution-based hybrid FEM for plane elasticity with special elements, Computational Mechanics 48 (2011) 515−528.

[16] H. Wang, Q.H. Qin, A fundamental solution-based finite element model for analyzing multilayer skin burn injury, Journal of Mechanics in Medicine and Biology 12 (2012) 1250027.

[17] H. Wang, Q.H. Qin, Numerical implementation of local effects due to two-dimensional discontinuous loads using special elements based on boundary integrals, Engineering Analysis With Boundary Elements 36 (2012) 1733−1745.

[18] H. Wang, Q.H. Qin, Boundary integral based graded element for elastic analysis of 2D functionally graded plates, European Journal of Mechanics − A: Solids 33 (2012) 12−23.

[19] H. Wang, Q.H. Qin, A new special element for stress concentration analysis of a plate with elliptical holes, Acta Mechanica 223 (2012) 1323−1340.

[20] H. Wang, Q.H. Qin, A new special coating/fiber element for analyzing effect of interface on thermal conductivity of composites, Applied Mathematics and Computation 268 (2015) 311−321.

[21] H. Wang, Q.H. Qin, Voronoi polygonal hybrid finite elements with boundary integrals for plane isotropic elastic problems, International Journal of Applied Mechanics 9 (2017) 1750031.

[22] H. Wang, Q.H. Qin, Y.P. Lei, Green's-function-based-finite element analysis of homogeneous fully plane anisotropic elastic bodies, Journal of Mechanical Science and Technology 31 (2017) 1305−1313.

[23] H. Wang, Q.H. Qin, Y. Xiao, Special n-sided Voronoi fiber/matrix elements for clustering thermal effect in natural-hemp-fiber-filled cement composites, International Journal of Heat and Mass Transfer 92 (2016) 228−235.

[24] H. Wang, Q.H. Qin, W. Yao, Improving accuracy of mode I stress intensity factor using fundamental solution based finite element model, Australian Journal of Mechanical Engineering 10 (2012) 41−52.

[25] H. Wang, X.J. Zhao, J.S. Wang, Interaction analysis of multiple coated fibers in cement composites by special n-sided interphase/fiber elements, Composites Science and Technology 118 (2015) 117−126.

[26] S.N. Atluri, S. Shen, The meshless local petrov-galerkin (mlpg) method: asimple & less-costly alternative to the finite element and boundary element methods, Computer Modeling in Engineering and Sciences 3 (2002) 11−51.

[27] G.R. Liu, Mesh Free Methods: Moving beyond the Finite Element Method, CRC Press, New York, 2003.

[28] E. Larsson, B. Fornberg, A numerical study of some radial basis functions based solution methods for elliptic PDEs, Computers and Mathematics With Applications 46 (2003) 891−902.

[29] M.A. Golberg, C.S. Chen, S.R. Karur, Improved multiquadric approximation for partial differential equations, Engineering Analysis With Boundary Elements 18 (1996) 9−17.

Part II

Applications of the meshless method

Chapter 4

Meshless analysis for thin beam bending problems

Chapter outline

4.1 Introduction

The mechanical response of slender structures is an important topic in engineering [1]. The bending theory of the Euler–Bernoulli beam has been described in Chapter 2, and it is clearly found that the bending problem of Euler–Bernoulli beam is a one-dimensional fourth-order boundary value problem [1,2]. For such problems, the classic finite element solutions have been widely determined by dividing beam structure into some beam elements for engineering applications [3,4]. More recently, numerical solutions based on collocation discretization of a beam have been obtained by meshless methods, e.g., the meshless local Petrov-Galerkin (MLPG) method, the local point interpolation method, the smoothed hydrodynamic particle (SPH) method, the radial basis function interpolation method, and the method of fundamental solutions in literature [5–9]. These meshless methods are uniquely simple and can provide highly accurate solutions that compete with those of finite elements without the difficulty of mesh connectivity [10].

In this chapter, the global strong-form meshless collocation formulation is presented to deal with one-dimensional fourth-order differential equation in Euler–Bernoulli beam bending problems, based on the one-dimensional

Methods of Fundamental Solutions in Solid Mechanics. https://doi.org/10.1016/B978-0-12-818283-3.00004-X
© 2019 Higher Education Press. Published by Elsevier Inc. All rights reserved.

polynomial basis functions and the radial basis functions (RBF). The one-dimensional polynomial basis functions are employed for constructing the exact homogeneous solutions, and the RBF are used for approximating particular solutions. The final satisfaction of boundary conditions can determine all unknowns in the meshless approximation. The effectiveness of the present meshless formulation is demonstrated through numerical examples of statically determinate or indeterminate beams under various loads and boundary conditions.

4.2 Solution procedures

Consider a thin beam bending under arbitrary load $q(x)$, as indicated in Fig. 4.1. The generalized one-dimensional governing equation of the Euler–Bernoulli beam in terms of four times differentiable variable w is recalled as

$$\frac{d^2 EI_y}{dx^2}\frac{d^2 w}{dx^2} + 2\frac{dEI_y}{dx}\frac{d^3 w}{dx^3} + EI_y\frac{d^4 w}{dx^4} = q(x) \tag{4.1}$$

which can be converted by means of the analog equation method [11] into

$$\frac{d^4 w}{dx^4} = \frac{1}{EI_y}\left[q(x) - \frac{d^2 EI_y}{dx^2}\frac{d^2 w}{dx^2} - 2\frac{dEI_y}{dx}\frac{d^3 w}{dx^3}\right] = b(x) \tag{4.2}$$

where $b(x)$ is a generalized fictitious source term that may be known if a uniform beam is considered or is an unknown function depending on the unknown deflection $w(x)$.

Eq. (4.2) indicates that the solution of Eq. (4.1) can be established by solving Eq. (4.1) under the same boundary conditions, provided that the fictitious load distribution $b(x)$ is first determined.

Due to the linearity of the biharmonic operator, the sought deflection solution $w(x)$ can be divided into two parts: the homogeneous part $w_h(x)$ and the particular part $w_p(x)$, which separately satisfy the following equations:

$$\frac{d^4 w_h}{dx^4} = 0 \tag{4.3}$$

and

$$\frac{d^4 w_p}{dx^4} = b(x) \tag{4.4}$$

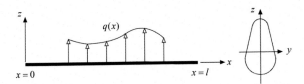

FIGURE 4.1 Thin beam bending under arbitrary load.

4.2.1 Homogeneous solution

Since the homogeneous Eq. (4.3) is in terms of the spatial variable x only, the homogeneous solution $w_h(x)$ can be obtained by means of the simple one-dimensional integral procedure:

$$w_h = c_0 + c_1 x + c_2 x^2 + c_3 x^3 = \begin{bmatrix} 1 & x & x^2 & x^3 \end{bmatrix} \begin{bmatrix} c_0 \\ c_1 \\ c_2 \\ c_3 \end{bmatrix} \tag{4.5}$$

where c_i $(i = 0, 1, 2, 3)$ are arbitrary integration constants to be determined from the specific boundary conditions.

Differentiating Eq. (4.5) we have

$$\frac{dw_h}{dx} = c_1 + 2c_2 x + 3c_3 x^2 \tag{4.6}$$

$$\frac{d^2 w_h}{dx^2} = 2c_2 + 6c_3 x \tag{4.7}$$

$$\frac{d^3 w_h}{dx^3} = 6c_3 \tag{4.8}$$

4.2.2 Particular solution

Based on the theory of RBF approximation [12,13], the fictitious load term $b(x)$ can be expressed as a linear combination of a set of RBF $\phi_j(r) = \phi(x, x_j)$ centered at x_j:

$$b(x) = \sum_{j=1}^{N_I} \alpha_j \phi_j(r) = \begin{bmatrix} \phi_1 & \phi_2 & \cdots & \phi_{N_I} \end{bmatrix} \begin{bmatrix} \alpha_1 \\ \alpha_2 \\ \vdots \\ \alpha_{N_I} \end{bmatrix} \tag{4.9}$$

where N_I is the number of interpolation points for RBF, as shown in Fig. 4.2, and

$$r = |x - x_j| \tag{4.10}$$

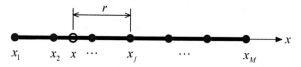

FIGURE 4.2 Interpolation points for radial basis function.

denotes the Euclidean distance between any point x and the given interpolating point x_j.

If we can find a set of particular solutions $\widehat{w}_j$ satisfying a relation as

$$\frac{d^4\widehat{w}_j}{dx^4} = \phi_j(r) \tag{4.11}$$

then it is reasonable to express the sought particular solution $w_p(x)$ in a similar form as Eq. (4.9):

$$w_p(x) = \sum_{j=1}^{N_I} \alpha_j \widehat{w}_j(r) = \begin{bmatrix} \widehat{w}_1 & \widehat{w}_2 & \cdots & \widehat{w}_{N_I} \end{bmatrix} \begin{bmatrix} \alpha_1 \\ \alpha_2 \\ \vdots \\ \alpha_{N_I} \end{bmatrix} \tag{4.12}$$

Currently, there are some RBFs available for Eq. (4.9). Here, the general linear spline RBF $\phi_j(r) = 1 + r$ is used in the study. Integrating Eq. (4.11) yields the corresponding set of particular solutions:

$$\widehat{w}_j(r) = \frac{r^4}{24} + \frac{r^5}{120} \tag{4.13}$$

Also, the derivatives of Eq. (4.13) can be written as

$$\frac{d\widehat{w}_j}{dx} = \left(\frac{r^2}{6} + \frac{r^3}{24}\right) r_j \tag{4.14}$$

$$\frac{d^2\widehat{w}_j}{dx^2} = \frac{r^2}{2} + \frac{r^3}{6} \tag{4.15}$$

$$\frac{d^3\widehat{w}_j}{dx^3} = \left(1 + \frac{r}{2}\right) r_j \tag{4.16}$$

where $r_j = x - x_j$.

4.2.3 Approximated full solution

Thus, the full expression of the sought solution $w(x)$ is obtained by totaling the homogeneous and particular parts:

$$w(x) = c_0 + c_1 x + c_2 x^2 + c_3 x^3 + \sum_{j=1}^{N_I} \alpha_j \widehat{w}_j(r) \tag{4.17}$$

In matrix notation, Eq. (4.17) is rewritten as

$$w(x) = \mathbf{p}^T \mathbf{c} + \mathbf{\Phi}^T \mathbf{\alpha} = \begin{bmatrix} \mathbf{p}^T & \mathbf{\Phi}^T \end{bmatrix} \begin{bmatrix} \mathbf{c} \\ \mathbf{\alpha} \end{bmatrix} = \mathbf{W}^T \mathbf{a} \tag{4.18}$$

where

$$\mathbf{p} = \begin{bmatrix} 1 & x & x^2 & x^3 \end{bmatrix}^T \tag{4.19}$$

$$\boldsymbol{\Phi} = \begin{bmatrix} \widehat{w}_1 & \widehat{w}_2 & \cdots & \widehat{w}_{N_I} \end{bmatrix}^T \tag{4.20}$$

$$\mathbf{c} = \begin{bmatrix} c_0 & c_1 & c_2 & c_3 \end{bmatrix}^T \tag{4.21}$$

$$\boldsymbol{\alpha} = \begin{bmatrix} \alpha_1 & \alpha_2 & \cdots & \alpha_{N_I} \end{bmatrix}^T \tag{4.22}$$

and

$$\mathbf{W} = [\, \mathbf{p}^T \quad \boldsymbol{\Phi}^T \,]^T, \qquad \mathbf{a} = \begin{bmatrix} \mathbf{c} \\ \boldsymbol{\alpha} \end{bmatrix} \tag{4.23}$$

4.2.4 Construction of solving equations

Correspondingly, the derivatives of Eq. (4.18) can be written as

$$\frac{d^n w(x)}{dx^n} = \frac{d^n \mathbf{W}^T}{dx^n} \mathbf{a}, \quad n = 1, 2, 3, 4 \tag{4.24}$$

with

$$\frac{d^n \mathbf{W}^T}{dx^n} = \begin{bmatrix} \dfrac{d^n \mathbf{p}^T}{dx^n} & \dfrac{d^n \boldsymbol{\Phi}^T}{dx^n} \end{bmatrix} \tag{4.25}$$

further, we have

$$\theta(x) = \frac{dw}{dx} = \frac{d\mathbf{W}^T}{dx} \mathbf{a}$$

$$M(x) = -EI_y \frac{d^2 w}{dx^2} = -EI_y \frac{d^2 \mathbf{W}^T}{dx^2} \mathbf{a}$$

$$Q(x) = -\frac{dEI_y}{dx} \frac{d^2 \mathbf{W}^T}{dx^2} \mathbf{a} - EI_y \frac{d^3 \mathbf{W}^T}{dx^3} \mathbf{a} = \left(-\frac{dEI_y}{dx} \frac{d^2 \mathbf{W}^T}{dx^2} - EI_y \frac{d^3 \mathbf{W}^T}{dx^3} \right) \mathbf{a}$$

$$\tag{4.26}$$

Substituting Eq. (4.24) into the governing Eq. (4.1) at the L interpolation points x_j ($j = 1 \rightarrow N_I$) produces

$$\left(\frac{d^2 EI_y}{dx^2} \frac{d^2 \mathbf{W}^T}{dx^2} + 2 \frac{dEI_y}{dx} \frac{d^3 \mathbf{W}^T}{dx^3} + EI_y \frac{d^4 \mathbf{W}^T}{dx^4} \right) \Bigg|_{x=x_j} \mathbf{a} = q(x_j), \quad j = 1 \rightarrow N_I$$

$$\tag{4.27}$$

Simultaneously, the satisfaction of the four boundary conditions at the two ends, i.e., $x = 0$ and $x = l$, gives

$$
\begin{bmatrix}
\mathbf{W}^{\mathrm{T}} \\[4pt]
\dfrac{d\mathbf{W}^{\mathrm{T}}}{dx} \\[8pt]
-EI_y\dfrac{d^2\mathbf{W}^{\mathrm{T}}}{dx^2} \\[8pt]
-\dfrac{dEI_y}{dx}\dfrac{d^2\mathbf{W}^{\mathrm{T}}}{dx^2} - EI_y\dfrac{d^3\mathbf{W}^{\mathrm{T}}}{dx^3}
\end{bmatrix}_{x=0,l} \mathbf{a} = 0 \qquad (4.28)
$$

Combining Eqs. (4.27) and (4.28), we would obtain a linear system of equations in abbreviated form,

$$
\mathbf{A}_{(N_I+4)\times(N_I+4)}\,\mathbf{a}_{(N_I+4)\times 1} = \mathbf{b}_{(N_I+4)\times 1} \qquad (4.29)
$$

which can be solved numerically to determine all unknowns in the vector $\mathbf{a}$. Then the deflection solution and the derived rotation, shear force, and moment at arbitrary point x in the beam can be evaluated using Eqs. (4.18) and (4.26).

4.2.5 Treatment of discontinuous loading

1. The discontinuously distributed loading

If the distributed loading is discontinuous at a point $x = x_0$, as indicated in Fig. 4.3, the mean value can be employed to restore the continuity in general. Here, the result is greatly improved by introducing a smoothed curve defined in a small range $[x_0 - \varepsilon, x_0 + \varepsilon]$ of length 2ε:

$$
q(x) = \frac{q_1 + q_2}{2} + \frac{q_2 - q_1}{2}\sin\frac{\pi(x - x_0)}{2\varepsilon}, \quad (x_0 - \varepsilon \le x \le x_0 + \varepsilon) \qquad (4.30)
$$

where q_1 and q_2 are values of the distributed load at the point x_0, and ε is a prescribed small value.

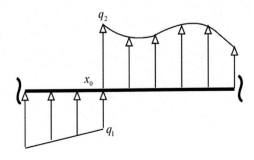

FIGURE 4.3 Illustration of discontinuously distributed loading.

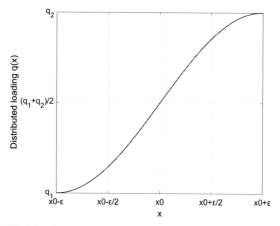

FIGURE 4.4 Continuous treatment of discontinuous distributed loading.

Fig. 4.4 shows the variation of Eq. (4.30) in the small range $[x_0 - \varepsilon, x_0 + \varepsilon]$, and it is clear that Eq. (4.30) can produce a smoothed variation around the point x_0, so the RBF interpolation can be successively implemented around the point x_0.

2. The concentrated force

For the case of concentrated force P at a point $x = x_0$, as displayed in Fig. 4.5, the generalized load $q(x)$ can be written as

$$q(x) = P\delta(x, x_0) \tag{4.31}$$

which causes a discontinuity at the point x_0.

To replace Eq. (4.31) in the computation, a bell-shaped continuous function defined in a small region of length 2ε is introduced (see Fig. 4.6), for instance,

$$q(x) = \frac{P}{2\varepsilon}\left[1 + \cos\frac{\pi(x - x_0)}{\varepsilon}\right], \quad (x_0 - \varepsilon \leq x \leq x_0 + \varepsilon) \tag{4.32}$$

so that

$$\int_{x_0-\varepsilon}^{x_0+\varepsilon} p(x)\mathrm{d}x = P \tag{4.33}$$

holds.

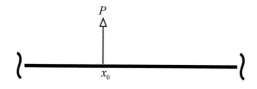

FIGURE 4.5 Illustration of discontinuous concentrated force.

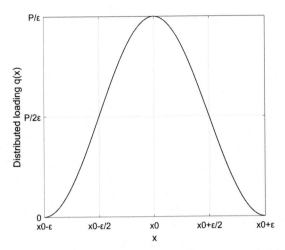

FIGURE 4.6 Continuous treatment of discontinuous concentrated force.

4.3 Results and discussion

To demonstrate the effectiveness of the proposed approach, three examples including two statically indeterminate beams and a statically determinate beam are taken into consideration. The first two tests are performed to demonstrate the ability of the present meshless approach for the case of a nonzero distributed force; the last test is performed for the case without distributed force. For the sake of simplicity, it is assumed that the beam cross-section is uniform along the axial direction of the beam and rectangular. It means that the moment of inertia of the cross-section about the y-axis remains constant in the computation. However, for a nonprismatic beam with the moment of inertia changing with respect to the spatial variable x, a similar solution procedure can be applied without any difficulty.

4.3.1 Statically indeterminate beam under uniformly distributed loading

A single statically indeterminate beam under the constant load $q(x) = -q_0$ is considered, as shown in Fig. 4.7. Its two ends are taken to be fixed at $x = 0$ and

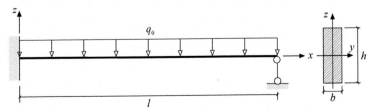

FIGURE 4.7 Statically indeterminate beam under uniformly distributed loading.

simply supported at $x = l$. The related boundary conditions are, respectively, $w(0) = 0$, $\theta(0) = 0$ and $w(l) = 0$, $M(l) = 0$.

In computation, the steel beam with length $l = 3$ m and rectangular cross-section with $b = 0.04$ m and $h = 0.1$ m is calculated under its weight $q_0 = 314$ N/m. The corresponding material constants are Young's modulus $E = 2.1 \times 10^{11}$ N/m^2 and Poisson's ration $v = 0.3$, respectively.

The corresponding exact solutions according to the Euler−Bernoulli beam theory are easily obtained as

$$w = \frac{-q_0}{EI_y}\left(-\frac{1}{24}x^4 + \frac{5l}{48}x^3 - \frac{l^2}{16}x^2\right)$$

$$\theta = \frac{-q_0}{EI_y}\left(-\frac{1}{6}x^3 + \frac{5l}{16}x^2 - \frac{l^2}{8}x\right)$$

$$M = q_0\left(-\frac{1}{2}x^2 + \frac{5l}{8}x - \frac{l^2}{8}\right)$$

$$Q = q_0\left(-x + \frac{5l}{8}\right)$$

$$(4.34)$$

in which the moment of inertial of the cross-section is

$$I_y = \frac{1}{3} \times 10^{-5} \text{m}^4 \qquad (4.35)$$

Using the proposed meshless approach with 10 uniformly distributed interpolation points, the numerical solutions of deflection, rotation, moment, and shear force are plotted in Figs. 4.8−4.11, from which it is clear that the proposed meshless approach can produce numerical results that agree well with the exact solutions.

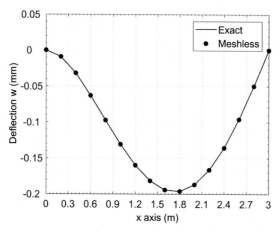

FIGURE 4.8 Distribution of deflection of the statically indeterminate beam under uniformly distributed loading.

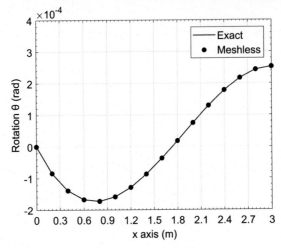

FIGURE 4.9 Distribution of rotation of the statically indeterminate beam under uniformly distributed loading.

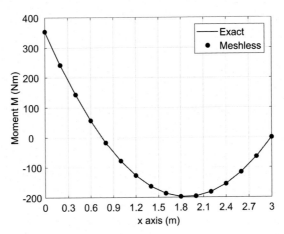

FIGURE 4.10 Distribution of moment of the statically indeterminate beam under uniformly distributed loading.

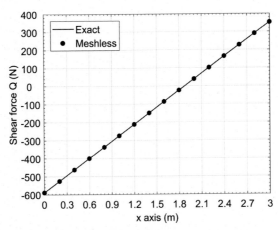

FIGURE 4.11 Distribution of shear force of the statically indeterminate beam under uniformly distributed loading.

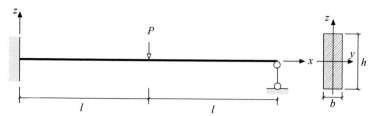

FIGURE 4.12 Statically indeterminate beam under middle-concentrated load.

4.3.2 Statically indeterminate beam under middle-concentrated load

Consider the statically indeterminate beam shown in Fig. 4.12. A concentrated force is applied at the middle of the beam. It is assumed that $E = 210$ GPa, $I_y = 60 \times 10^{-6}$ m^4, $P = 20$ kN, and $l = 2$ m. The related boundary conditions are $w(0) = 0$, $\theta(0) = 0$ and $w(2l) = 0$, $M(2l) = 0$.

With 16 interpolation points, the numerical solutions of deflection, rotation, moment, and shear force can be obtained and are tabulated in Table 4.1, in which the finite element results are also provided for comparison. It is found that the proposed meshless approach can produce almost the same numerical results as those from the finite element method (FEM) [3], but the time-consuming task for mesh connectivity in FEM is fully avoided. Besides, the results in Figs. 4.13–4.16 show the present meshless approach can generate smoothed deflection and rotation curves, while the internal shear force Q retains the sudden change at the midpoint of the beam, and the moment M retains the derivative discontinuity, as expected.

4.3.3 Cantilever beam with end-concentrated load

In the final test, let us consider a cantilever beam shown in Fig. 4.17. A concentrated force is applied at the right end of the beam. The material property of the beam is $E = 210$ GPa. The specific boundary conditions are $w(0) = 0$, $\theta(0) = 0$ and $Q(l) = -P$, $M(l) = 0$. For that beam, the distributed load $P(x)$ in the governing Eq. (4.1) is zero. This test is taken to investigate the ability of the present meshless approach to treat the special case without transverse distributed force.

The exact solutions that can be found in most textbooks of mechanics of materials are

$$w(x) = \frac{1}{3} \frac{Pl^3}{EI_y} \left[-\frac{1}{2} \left(\frac{l-x}{l} \right)^3 + \frac{3}{2} \left(\frac{l-x}{l} \right) - 1 \right]$$

$$\theta(x) = \frac{1}{3} \frac{Pl^3}{EI_y} \left[\frac{3}{2l} \left(\frac{l-x}{l} \right)^2 - \frac{3}{2l} \right] \tag{4.36}$$

$$M(x) = P(l-x)$$

$$Q(x) = -P$$

TABLE 4.1 Comparison of results from the present meshless method and the FEM.

Location (m)	w (mm)		θ (rad)		M (kNm)		Q (kN)	
	FEM	Meshless	FEM	Meshless	FEM	Meshless	FEM	Meshless
x = 0	0.0000	0.0000	0.0000	0.0000		15.0000	−13.748	−13.750
x = 2	−0.9259	−0.9259	−0.1984	−0.1984		−12.4967	/	/
x = 4	0.0000	0.0000	0.7937	0.7937		0.0000	6.252	6.250

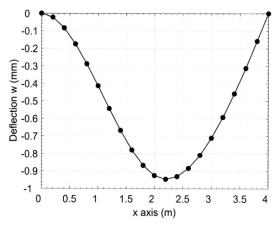

FIGURE 4.13 Distribution of deflection of statically indeterminate beam under concentrated load.

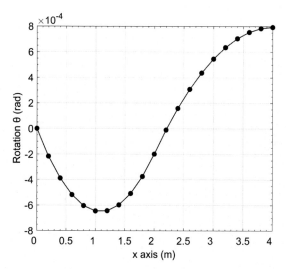

FIGURE 4.14 Distribution of rotation angle of statically indeterminate beam under concentrated load.

In the computation, it is assumed that $I_y = 60 \times 10^{-6}$ m^4, $P = 20$ kN, and $l = 4$ m. The 10 interpolation points are uniformly placed along the beam. Figs. 4.18–4.21 plot the variations of deflection, rotation, moment, and shear force, respectively, along the beam, from which it is clear that the proposed meshless approach can produce results in agreement with the exact solutions.

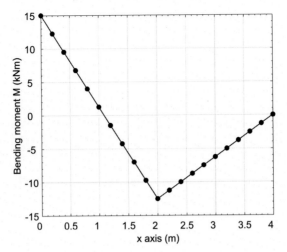

FIGURE 4.15 Distribution of bending moment of statically indeterminate beam under concentrated load.

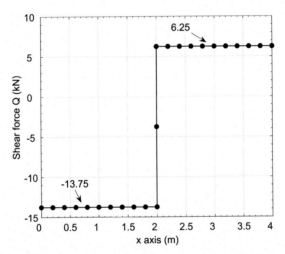

FIGURE 4.16 Distribution of shear force of statically indeterminate beam under concentrated load.

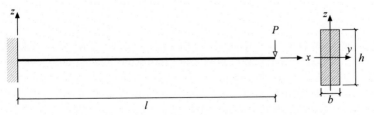

FIGURE 4.17 Cantilever beam with tip-concentrated load.

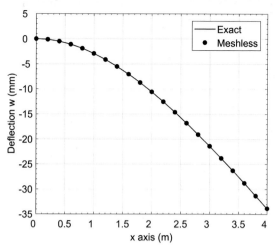

FIGURE 4.18 Distribution of deflection of cantilever beam with tip-concentrated load.

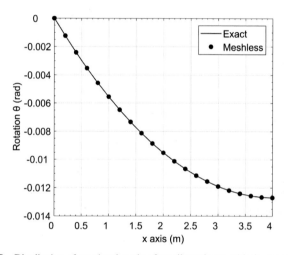

FIGURE 4.19 Distribution of rotational angle of cantilever beam with tip-concentrated load.

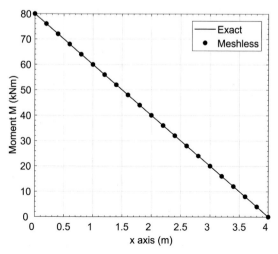

FIGURE 4.20 Distribution of bending moment of cantilever beam with tip-concentrated load.

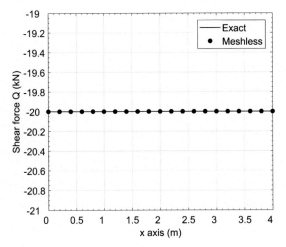

FIGURE 4.21 Distribution of shear force of cantilever beam with tip-concentrated load.

4.4 Remarks

A meshless solution strategy is presented in this chapter to solve the one-dimensional fourth-order boundary value problems of isotropic homogeneous Euler–Bernoulli thin beams. In the present meshless method, the analog equation method is firstly employed to convert the generalized fourth-order differential equation into an equivalent linear one with a fictitious right-handed term. Then, the full solution of deflection in the equivalent fourth-order equation is divided into two parts: the homogeneous solution and the particular solution, to deal with arbitrarily distributed transverse loadings. The homogeneous solution is expressed as a polynomial expression, and the particular solution is approximated by the radial basis function interpolation. Finally the full solution and its derivatives are incorporated to satisfy the real governing equation and the specific boundary conditions.

Numerical examples of static analysis for thin beams under various loads and boundary conditions are analyzed to demonstrate the effectiveness of the present approach. It is found that the present meshless solving strategy is easily implemented and very flexible for static analysis of thin beams. Also, the present approach is easily extended to solve more complex bending problems such as functionally graded beams, taped beams, vibrating beams, beams on elastic foundation, and microbeams in a micro-electro-mechanical system (MEMS) [8,9,14–16].

References

[1] L.H. Donnell, Beams, Plates, and Shells, McGraw-Hill, New York, 1976.
[2] Z.H. Wang, X.H. Wang, G.D. Xu, S. Cheng, T. Zeng, Free vibration of two-directional functionally graded beams, Composite Structures 135 (2016) 191–198.

[3] K.J. Bathe, Finite Element Procedures, Prentice-Hall, Inc., New Jersey, 1996.

[4] N.T. Nguyen, N.I. Kim, J. Lee, Mixed finite element analysis of nonlocal Euler–Bernoulli nanobeams, Finite Elements in Analysis and Design 106 (2015) 65–72.

[5] Y.T. Gu, G.R. Liu, A local point interpolation method for static and dynamic analysis of thin beams, Computer Methods in Applied Mechanics and Engineering 190 (2001) 5515–5528.

[6] I.S. Raju, D.R. Phillips, T. Krishnamurthy, A radial basis function approach in the meshless local Petrov-Galerkin method for Euler-Bernoulli beam problems, Computational Mechanics 34 (2004) 464–474.

[7] J.R. Xiao, M.A. McCarthy, Meshless analysis of the obstacle problem for beams by the MLPG method and subdomain variational formulations, European Journal of Mechanics – A: Solids 22 (2003) 385–399.

[8] H.J. Al-Gahtani, F.M. Mukhtar, RBF-based meshless method for the free vibration of beams on elastic foundations, Applied Mathematics and Computation 249 (2014) 198–208.

[9] J. Lin, J. Li, Y. Guan, G. Zhao, H. Naceur, D. Coutellier, Geometrically nonlinear bending analysis of functionally graded beam with variable thickness by a meshless method, Composite Structures 189 (2018) 239–246.

[10] G.R. Liu, Mesh Free Methods: Moving beyond the Finite Element Method, CRC Press, New York, 2003.

[11] J.T. Katsikadelis, The Boundary Element Method for Engineers and Scientists: Theory and Applications, Elsevier, 2016.

[12] M.D. Buhmann, Radial basis functions, Acta Numerica 9 (2003) 1–38.

[13] R. Schaback, Error estimates and condition numbers for radial basis function interpolation, Advances in Computational Mathematics 3 (1995) 251–264.

[14] H. Li, Q.X. Wang, K.Y. Lam, A variation of local point interpolation method (vLPIM) for analysis of microelectromechanical systems (MEMS) device, Engineering Analysis with Boundary Elements 28 (2004) 1261–1270.

[15] Y. Zhao, Y. Huang, M. Guo, A novel approach for free vibration of axially functionally graded beams with non-uniform cross-section based on Chebyshev polynomials theory, Composite Structures 168 (2017) 277–284.

[16] J.J. Allen, Micro Electro Mechanical System Design, CRC Press, New York, 2005.

Chapter 5

Meshless analysis for thin plate bending problems

Chapter outline

5.1 Introduction

As a type of thin plane structures, thin plate theory based on the Kirchhoff hypothesis have been widely studied in practice, spanning from traditional structural engineering to recently developed micro-electro-mechanical systems, in which thin plate—shaped conductors are usually adopted [1—4]. Such a Kirchhoff plate model is typically governed by a fourth-order partial differential equation with transverse displacement variable [5]. For thin plate bending problems with arbitrary transverse loads and boundary conditions, the analysis is usually carried out by numerical approaches, instead of theoretical procedures. For instance, the boundary element method (BEM) [6,7], the hybrid finite element method [8,9], and the finite element method (FEM) [10] were developed to analyze various thin plate deformation problems, respectively. Additionally, meshless methods such as the meshless local Petrov-Galerkin method, the local boundary integral equation method, and the method of fundamental solution (MFS) were established during the past decades for thin plate bending problems [11—16] because of their flexibility and wide applicability over the conventional mesh-dependent methods like FEM and BEM.

In this chapter, we focus on numerical solutions of the classical thin plate model under transverse loads by the extended MFSs. Based on the characteristics of the governing equation of thin plate, the analog equation method is

Methods of Fundamental Solutions in Solid Mechanics. https://doi.org/10.1016/B978-0-12-818283-3.00005-1
127

firstly used to convert the original fourth-order partial differential governing equation into an equivalent one, which usually has a simpler expression, then the MFS and the radial basis function interpolation are respectively employed to yield the related homogeneous and particular solutions. The strong-form satisfaction of the governing equation and boundary conditions at interior and boundary collocations can be used to determine all unknowns in the solving system.

5.2 Fundamental solutions for thin plate bending

For a thin plate bending problem, a two-dimensional fourth-order partial differential equation is employed to describe its transverse deformation, as depicted in Section 2.5:

$$D\nabla^4 w(\mathbf{x}) = p(\mathbf{x}) \tag{5.1}$$

where $w(\mathbf{x})$ denotes the lateral displacement (deflection) of interest at an arbitrary point $\mathbf{x} = (x, y) \in \Omega \subset \mathbb{R}^2$, and normally it is fourth-order differentiable, $p(\mathbf{x})$ is the applied transverse load, and ∇^4 is the biharmonic differential operator defined by

$$\nabla^4 = \frac{\partial^4}{\partial x^4} + 2\frac{\partial^4}{\partial x^2 \partial y^2} + \frac{\partial^4}{\partial y^4} \tag{5.2}$$

In Eq. (5.1), the flexural rigidity D is given by

$$D = \frac{Eh^3}{12(1 - v^2)} \tag{5.3}$$

where E is the Young's modulus of material, v is the Poisson's ratio, and h is the plate thickness.

Besides, if the thin plate is placed on an elastic Winkler foundation, the governing Eq. (5.1) can be extended to include the effect of the foundation

$$D\nabla^4 w(\mathbf{x}) + k_w w(\mathbf{x}) = p(\mathbf{x}) \tag{5.4}$$

where k_w is the coefficient of the elastic foundation.

Following the treatment of analog equation method presented in Ref. [6], the fourth-order plate bending Eqs. (5.1) and (5.4) can be rewritten in a unified form as

$$D\nabla^4 w(\mathbf{x}) = \widetilde{p}(\mathbf{x}) \tag{5.5}$$

where

$$\widetilde{p}(\mathbf{x}) = \begin{cases} p(\mathbf{x}) \\ p(\mathbf{x}) - k_w w(\mathbf{x}) \end{cases} \tag{5.6}$$

for general thin plate and thin plate on elastic foundation, respectively.

The boundary conditions on a smoothed boundary of thin plate domain can be of any combination of clamped, simply supported, and free boundary conditions, as described in Section 2.6, and only two boundary conditions are considered at each boundary point because the governing plate equation is a fourth-order partial differential equation. Here, for the sake of convenience, we rewrite the boundary conditions in general form as

$$\begin{aligned} w &= \overline{w}_0 \\ \theta_n &= \overline{\theta}_n \\ V_n &= \overline{V}_n \\ M_n &= \overline{M}_n \end{aligned} \tag{5.7}$$

where θ_n, V_n, and M_n represent the normal rotation, the effective shear force, and the bending moment perpendicular to the normal direction n of the boundary Γ, respectively. They can be expressed in terms of the lateral deflection w, as given in Eq. (2.85). $\overline{w}$, $\overline{\theta}_n$, $\overline{V}_n$, and $\overline{M}_n$ are prescribed values, respectively.

For the thin plate bending problem, two sets of fundamental solutions given as follows are involved in the meshless analysis.

On one hand, for the biharmonic operator that appeared in Eq. (5.5), the corresponding fundamental solution, also known as Almansi's fundamental solution, is required to satisfy

$$\nabla^4 w_1^*(\mathbf{x}, \mathbf{y}) + \delta(\mathbf{x}, \mathbf{y}) = 0 \quad \forall \mathbf{x}, \mathbf{y} \in \mathbb{R}^2 \tag{5.8}$$

which represents the deflection response at a field point $\mathbf{x}$ in an infinitely free plate caused by a point force at the source point $\mathbf{y}$.

Because we are dealing with an infinite domain problem, there is no disturbance from the boundary. Moreover, the excitation with the Dirac delta function $\delta(\mathbf{x}, \mathbf{y})$ is radial symmetric, which can be written in polar coordinates as [17]

$$\delta(\mathbf{x}, \mathbf{y}) = \frac{\delta(r)}{2\pi r} \tag{5.9}$$

with

$$r = \|\mathbf{x} - \mathbf{y}\| \tag{5.10}$$

Thus, it implies that the fundamental solution w_1^* we are looking for is naturally radial symmetric too. To derive such "radial" fundamental solution, the radial symmetry of biharmonic operator is considered. For such special radial-only case, the biharmonic operator in Eq. (5.8) in two-dimensional polar coordinates can be written as follows

$$\nabla^4 = \frac{1}{r} \frac{d}{dr} \left(r \frac{d}{dr} \left(\frac{1}{r} \frac{d}{dr} \left(r \frac{d}{dr} \right) \right) \right) \tag{5.11}$$

Therefore, one way to solve Eq. (5.8) for w_1^* is to integrate Eq. (5.11). As a result, we have Ref. [2]

$$w_1^*(\mathbf{x}, \mathbf{y}) = -\frac{1}{8\pi} r^2 \ln r \tag{5.12}$$

On the other hand, the general solution of the homogeneous biharmonic equation

$$\nabla^4 w(\mathbf{x}) = 0 \tag{5.13}$$

can be expressed as

$$w(\mathbf{x}) = A(\mathbf{x}) + r^2 B(\mathbf{x}) \tag{5.14}$$

where A and B are two independent functions satisfying the Laplace equation, respectively,

$$\nabla^2 A(\mathbf{x}) = 0, \quad \nabla^2 B(\mathbf{x}) = 0 \tag{5.15}$$

Thus, the fundamental solution of the Laplacian operator is also required to be the second kernel in the thin plate bending problem to form the complete fundamental solution. Correspondingly, the second fundamental solution can be obtained by solving

$$\nabla^2 w_2^*(\mathbf{x}, \mathbf{y}) + \delta(\mathbf{x}, \mathbf{y}) = 0 \quad \forall \mathbf{x}, \mathbf{y} \in \mathbb{R}^2 \tag{5.16}$$

Similarly, due to the radial symmetric property of the function w_2^* in polar coordinates, the Laplace operator can be written as

$$\nabla^2 = \frac{1}{r} \frac{\mathrm{d}}{\mathrm{d}r} \left(r \frac{\mathrm{d}}{\mathrm{d}r} \right) = \frac{\mathrm{d}^2}{\mathrm{d}r^2} + \frac{1}{r} \frac{\mathrm{d}}{\mathrm{d}r} \tag{5.17}$$

The solution of Eq. (5.16) can be obtained by integrating Eq. (5.17) as

$$w_2^*(\mathbf{x}, \mathbf{y}) = -\frac{1}{2\pi} \ln r \tag{5.18}$$

Fig. 5.1 displays the variations of the solutions w_1^* and w_2^* when a unit transverse point force is applied at the origin of Cartesian coordinate system, that is, $\mathbf{y} = \{0, 0\}$, and it is observed that both the solutions w_1^* and w_2^* are radial functions dependent on r only. In addition, the singularity of the fundamental solutions is represented by the kernel function $\ln r$.

5.3 Solutions procedure for thin plate bending

Considering the linearity of fourth-order differential operator ∇^4, we divide the solution $w(\mathbf{x})$ into two parts: homogeneous solution $w_h(\mathbf{x})$ and particular solution $w_p(\mathbf{x})$, which satisfy respectively

$$\nabla^4 w_h(\mathbf{x}) = 0 \tag{5.19}$$

and

$$\nabla^4 w_p(\mathbf{x}) = \tilde{p}(\mathbf{x}) \tag{5.20}$$

(A)

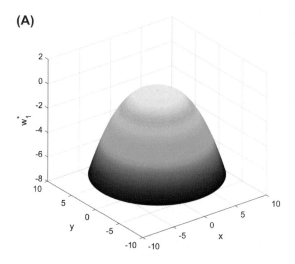

(B)

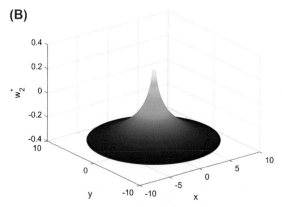

FIGURE 5.1 Variations of fundamental solutions w_1^* and w_2^* for $\mathbf{y} = (0, 0)$.

5.3.1 Particular solution

To obtain the particular solution related to the arbitrarily formed load $\widetilde{p}(\mathbf{x})$, the radial basis function is used to approximate both the load function and the particular solution. To do this, two approximations with similar form are introduced as follows:

$$\widetilde{p}(\mathbf{x}) \approx \sum_{j=1}^{N_I} \alpha_j \phi_j(\mathbf{x}) \tag{5.21}$$

$$w_p(\mathbf{x}) = \sum_{j=1}^{N_I} \alpha_j \Phi_j(\mathbf{x}) \tag{5.22}$$

where $\phi_j(\mathbf{x})$ and $\Phi_j(\mathbf{x})$ ($j = 1, 2,..., N_I$) respectively represent the radial basis function and the corresponding particular solution kernel, and N_I is the number of interpolation points in the plate domain.

TABLE 5.1 Particular solutions related to conical and TPS radial basis function (RBF) for the biharmonic operator.

	Conical RBF	Thin plate spline RBF
ϕ	r^{2n-1}	$r^{2n}\ln r$
Φ	$\dfrac{r^{2n+3}}{(2n+1)^2(2n+3)^2}$	$\dfrac{r^{2n+4}}{16(n+1)^2(n+2)^2}\left[\ln r - \dfrac{2n+3}{(n+1)(n+2)}\right]$

Eqs. (5.21) and (5.22) can be rewritten in matrix form as

$$\tilde{p}(\mathbf{x}) = \mathbf{f}\boldsymbol{\alpha} \tag{5.23}$$

$$w_p(\mathbf{x}) = \boldsymbol{\Phi}\boldsymbol{\alpha} \tag{5.24}$$

where

$$\boldsymbol{\alpha} = \begin{bmatrix} \alpha_1 & \alpha_2 & \cdots & \alpha_{N_I} \end{bmatrix}^{\mathrm{T}} \tag{5.25}$$

$$\mathbf{f} = \begin{bmatrix} \phi_1(\mathbf{x}) & \phi_2(\mathbf{x}) & \cdots & \phi_{N_I}(\mathbf{x}) \end{bmatrix} \tag{5.26}$$

$$\boldsymbol{\Phi} = \begin{bmatrix} \Phi_1(\mathbf{x}) & \Phi_2(\mathbf{x}) & \cdots & \Phi_{N_I}(\mathbf{x}) \end{bmatrix} \tag{5.27}$$

Substituting Eqs. (5.21) and (5.22) into Eq. (5.20) yields the following connecting equation between $\phi_j(\mathbf{x})$ and $\Phi_j(\mathbf{x})$:

$$\nabla^4 \Phi(r_j) = \phi(r_j) \tag{5.28}$$

where $\phi(r_j) = \phi_j(\mathbf{x}) = \phi(\mathbf{x}, \mathbf{x}_j)$, $\Phi(r_j) = \Phi_j(\mathbf{x}) = \Phi(\mathbf{x}, \mathbf{x}_j)$, and $r_j = \|\mathbf{x} - \mathbf{x}_j\|$.

It is clear that once a radial basis function ϕ is given, integrating Eq. (5.28) with respect to polar coordinates can generate the radial particular solution kernel Φ, due to the radial symmetry property of the biharmonic operator as given in Eq. (5.11). In Table 5.1, two types of radial basis functions and the related particular solution kernels are tabulated for late use.

5.3.2 Homogeneous solution

Since for a well-posed thin plate bending problem there are two known and two unknown boundary conditions at each point on the boundary, we can combine the fundamental solutions of biharmonic operator and Laplace operator to fulfill the character of the boundary conditions; that is, the homogeneous solution at field point $\mathbf{x}$ in the computing domain can be written as

$$w_h(\mathbf{x}) = \sum_{i=1}^{N_s} \left[\varphi_{1i} w_1^*(\mathbf{x}, \mathbf{y}_i) + \varphi_{2i} w_2^*(\mathbf{x}, \mathbf{y}_i) \right], \quad \mathbf{x} \in \Omega, \ \mathbf{y}_i \notin \Omega \tag{5.29}$$

where N_s is the number of source points outside the computational domain, and φ_{1i} and φ_{2i} are coefficients to be determined.

Eq. (5.29) can be written in matrix form as

$$w_h(\mathbf{x}) = \sum_{i=1}^{N_s} \mathbf{w}_i^*(\mathbf{x})\boldsymbol{\varphi}_i = \mathbf{w}^*(\mathbf{x})\boldsymbol{\varphi}, \quad \mathbf{x} \in \Omega, \ \mathbf{y}_i \notin \Omega \tag{5.30}$$

where

$$\mathbf{w}^*(\mathbf{x}) = \begin{bmatrix} \mathbf{w}_1^*(\mathbf{x}) & \mathbf{w}_2^*(\mathbf{x}) & \cdots & \mathbf{w}_{N_s}^*(\mathbf{x}) \end{bmatrix} \tag{5.31}$$

$$\boldsymbol{\varphi} = \begin{bmatrix} \boldsymbol{\varphi}_1 & \boldsymbol{\varphi}_2 & \cdots & \boldsymbol{\varphi}_{N_s} \end{bmatrix}^{\mathrm{T}} \tag{5.32}$$

with

$$\mathbf{w}_i^*(\mathbf{x}) = \begin{bmatrix} w_1^*(\mathbf{x}, \mathbf{y}_i) & w_2^*(\mathbf{x}, \mathbf{y}_i) \end{bmatrix}, \quad \boldsymbol{\varphi}_i = \begin{bmatrix} \varphi_{1i} \\ \varphi_{2i} \end{bmatrix} \tag{5.33}$$

Obviously, the fundamental solution w_1^* and w_2^* is not singular due to the fact that $\mathbf{x} \neq \mathbf{y}_i$ in the MFS, so the approximated solution (5.29) can satisfy analytically the homogeneous Eq. (5.19).

5.3.3 Approximated full solution

Making use of the particular solution and the homogeneous solution obtained before, we obtain

$$w(\mathbf{x}) = \sum_{i=1}^{N_s} \left[\phi_{1i} w_1^*(\mathbf{x}, \mathbf{y}_i) + \phi_{2i} w_2^*(\mathbf{x}, \mathbf{y}_i) \right] + \sum_{j=1}^{N_l} \alpha_j \Phi_j(\mathbf{x}) \tag{5.34}$$

It can be written in matrix form as

$$w(\mathbf{x}) = \mathbf{w}^*(\mathbf{x})\boldsymbol{\varphi} + \boldsymbol{\Phi}(\mathbf{x})\boldsymbol{\alpha} = \mathbf{W}(\mathbf{x})\boldsymbol{\beta} \tag{5.35}$$

where

$$\mathbf{W}(\mathbf{x}) = \begin{bmatrix} \mathbf{w}^*(\mathbf{x}) & \boldsymbol{\Phi}(\mathbf{x}) \end{bmatrix}, \quad \boldsymbol{\beta} = \begin{bmatrix} \boldsymbol{\varphi} \\ \boldsymbol{\alpha} \end{bmatrix} \tag{5.36}$$

Subsequently, the rotation, effective shear force and bending moment can be derived by Eq. (2.85) as follows:

$$\begin{aligned}
\theta_n &= w_{,i} n_i = L_{\theta_n} \mathbf{W}(\mathbf{x})\boldsymbol{\beta} \\
V_n &= -D\left[w_{,ijj} n_i + (1-v) w_{,ijk} n_i t_j t_k \right] = L_{V_n} \mathbf{W}(\mathbf{x})\boldsymbol{\beta} \\
M_n &= -D\left[v \partial w_{,kk} + (1-v)\partial w_{,kl} n_k n_l \right] = L_{M_n} \mathbf{W}(\mathbf{x})\boldsymbol{\beta}
\end{aligned} \tag{5.37}$$

where

$$\begin{aligned}
L_{\theta_n} &= \partial_{,i} n_i \\
L_{V_n} &= -D\left[\partial_{,ijj} n_i + (1-v)\partial_{,ijk} n_i t_j t_k \right] \\
L_{M_n} &= -D\left[v \partial_{,kk} + (1-v)\partial_{,kl} n_k n_l \right]
\end{aligned} \tag{5.38}$$

are partial differential operators related to boundary physical quantities in terms of the primary dependent variable w, respectively.

5.3.4 Construction of solving equations

The unknown vector $\boldsymbol{\beta}$ can be determined by substituting Eq. (5.35) into the original governing Eq. (5.5) at N_I interpolation points and under the specific boundary conditions (5.7) at N_B ($=N_s$ usually) boundary collocations, that is,

$$
\begin{bmatrix} \dfrac{L_{\nabla^4}\mathbf{W}}{\mathbf{W}} \\ L_{\theta_n}\mathbf{W} \\ L_{M_n}\mathbf{W} \\ L_{V_n}\mathbf{W} \end{bmatrix} \boldsymbol{\beta} = \begin{bmatrix} \dfrac{p}{\overline{w}} \\ \overline{\theta}_n \\ \overline{M}_n \\ \overline{V}_n \end{bmatrix} \tag{5.39}
$$

where L_{∇^4} is the partial differential operator of the governing equation in terms of the primary dependent variable w. For example, in the case of thin plate on an elastic foundation with simply supported boundary conditions, which involve the specific deflection and rotation constraints only along the plate boundary, we have

$$
L_{\nabla^4} = D\nabla^4 + k_w \tag{5.40}
$$

and

$$
\begin{bmatrix} \dfrac{L_{\nabla^4}\mathbf{W}}{\mathbf{W}} \\ L_{\theta_n}\mathbf{W} \end{bmatrix} \boldsymbol{\beta} = \begin{bmatrix} \dfrac{p}{\overline{w}} \\ \overline{\theta}_n \end{bmatrix} \tag{5.41}
$$

Therefore, the substitution of Eq. (5.35) into Eq. (5.41) yields the following solving system of linear equations

$$
\begin{bmatrix} L_{\nabla^4}\mathbf{W}(\mathbf{x}_1) \\ \vdots \\ L_{\nabla^4}\mathbf{W}(\mathbf{x}_{N_I}) \\ \hline L_w\mathbf{W}(\mathbf{x}_{b1}) \\ \vdots \\ L_w\mathbf{W}(\mathbf{x}_{bN_B}) \\ L_{\theta_n}\mathbf{W}(\mathbf{x}_{b1}) \\ \vdots \\ L_{\theta_n}\mathbf{W}(\mathbf{x}_{bN_B}) \end{bmatrix} \boldsymbol{\beta} = \begin{bmatrix} p(\mathbf{x}_1) \\ \vdots \\ p(\mathbf{x}_{N_I}) \\ \hline \overline{w}(\mathbf{x}_{b1}) \\ \vdots \\ \overline{w}(\mathbf{x}_{bN_B}) \\ \overline{\theta}_n(\mathbf{x}_{b1}) \\ \vdots \\ \overline{\theta}_n(\mathbf{x}_{bN_B}) \end{bmatrix} \tag{5.42}
$$

where $\mathbf{x}_i$ ($i = 1 \rightarrow N_I$) and $\mathbf{x}_{bi}$ ($i = 1 \rightarrow N_B$) are the interpolation points in the computing domain and the boundary collocations on the boundary of the computing domain, respectively. Then the deflection solution and the derived rotation, shear force, and moment at arbitrary point $\mathbf{x}$ in the plate can be evaluated using Eqs. (5.35) and (5.37).

5.4 Results and discussion

In this section, two examples are considered to investigate the performance of the proposed algorithm. The first example appertains to a general thin plate bending problem, and it is designed to demonstrate the convergence, stability, and feasibility. In the second example, a typical thin plate bending is placed on a Winkler elastic foundation. To provide a more quantitative understanding of results, the average relative error of the deflection w (Arerr(w)) defined in Eq. (1.51) is evaluated at all test points.

5.4.1 Square plate with simple-supported edges

Consider a square plate subjected to uniformly distributed load q_0, as shown in Fig. 5.2. The four edges of the plate are simply supported, which means that $w = 0$ and $\theta_n = 0$ are enforced along all boundary edges.

The available analytical solution of deflection for this problem is given by

$$w(x_1, x_2) = \sum_{m=1}^{\infty} \sum_{n=1}^{\infty} A_{mn} \sin\frac{m\pi x_1}{a} \sin\frac{n\pi x_2}{a} \tag{5.43}$$

with

$$A_{mn} = \frac{q_{mn}}{D\pi^4 \left(\dfrac{m^2 + n^2}{a^2}\right)^2}$$

$$q_{mn} = \frac{4q_0[-1 + \cos(m\pi)][-1 + \cos(n\pi)]}{mn\pi^2}$$

In the computation, the related parameters are set as $a = 1$ m, $h = 0.01$ m, $q_0 = 1$ kN/m^2, $E = 2.1 \times 10^4$ MPa, and $\nu = 0.3$. With these values, the analytical distribution of normalized deflection $wD/(q_0 a^4)$ is plotted in Fig. 5.3 for reference.

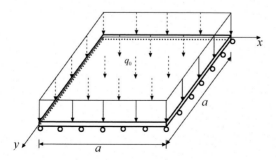

FIGURE 5.2 Square plate with all simply supported edges subjected to uniformly distributed load.

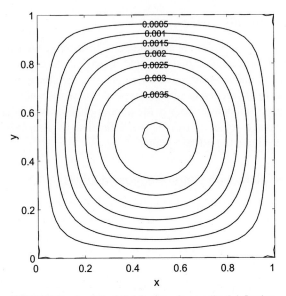

FIGURE 5.3   Analytical distribution of normalized deflection.

Firstly, the convergent behavior of the present meshless method is investigated. When 64 interior collocations are used for RBF interpolation, the variations of the error Arerr(w) are plotted in Figs. 5.4 and 5.5 with increasing boundary collocations N_s. The error Arerr(w) is evaluated by 441 test points

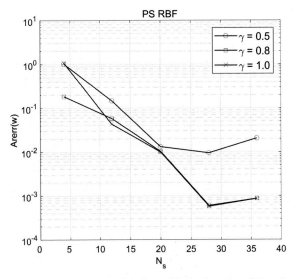

FIGURE 5.4   Convergence of the power spline with the increase of boundary collocations ($N_I = 64$).

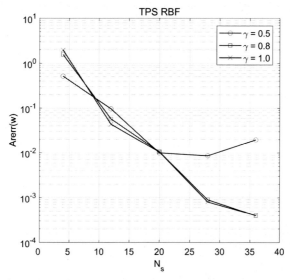

FIGURE 5.5 Convergence of the thin plate spline with the increase of boundary collocations ($N_I = 64$).

uniformly dispersed in the domain. Results in Figs. 5.4 and 5.5 show that the TPS-type radial basis function (RBF) is more stable than the PS-type RBF, especially for larger N_s. At the same time, the results affected by different values of the parameter γ that represents the distance of source points to the real physical boundary are also presented in Figs. 5.4 and 5.5. It is found that the location of source points affects the convergence of numerical solutions, and the larger γ leads to better accuracy and convergence. Accordingly, the TPS-type RBF and $\gamma = 0.8$ are chosen in the computation following. The convergence of the numerical solutions is investigated additionally for different numbers of boundary collocations. It is observed from Fig. 5.6 that a good convergence is achieved with the increase of N_I. All these results show a good convergence and accuracy of the proposed algorithm.

Finally, the distribution of deflection in the plate and the distributions of rotation angles, moments, and shear forces along $y = 0.5$ m under the conditions of $N_S = 36$, $N_I = 64$, and $\gamma = 0.8$ are plotted in Figs. 5.7–5.10, in which a good agreement between numerical results from the present method and analytical solutions is achieved.

5.4.2 Square plate on a winkler elastic foundation

In the second example, we take the same square plate in the previous example but place it on a Winkler foundation. The same boundary conditions and the transverse uniformly distributed load are taken into consideration as well. The coefficient of the Winkler foundation is set as $k_w = 4.9 \times 10^7$ N/m^3.

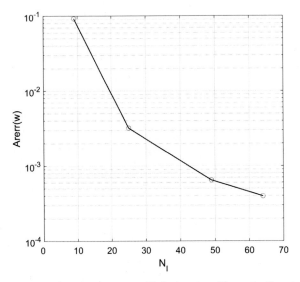

FIGURE 5.6 Demonstration of convergence with the increase of internal collocations ($N_S = 36$).

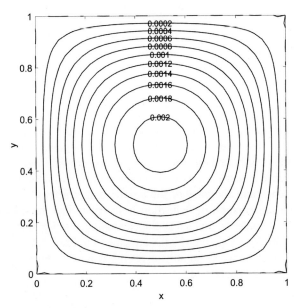

FIGURE 5.7 Numerical results of normalized deflection in the plate.

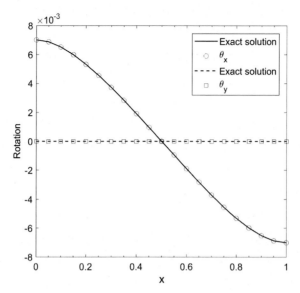

FIGURE 5.8 Distribution of rotation angle along the line $y = 0.5$ m.

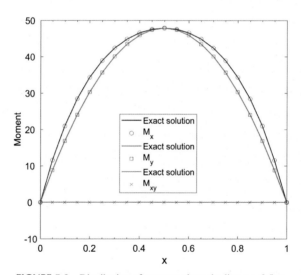

FIGURE 5.9 Distribution of moment along the line $y = 0.5$ m.

In this case, the analytical solution of deflection has same expression as that of Eq. (5.43), except for

$$A_{mn} = \frac{q_{mn}}{D\pi^4 \left(\dfrac{m^2 + n^2}{a^2}\right)^2 + k_w}$$

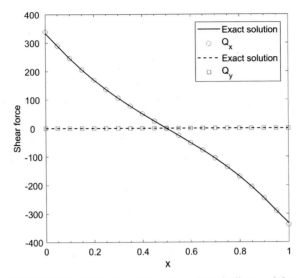

FIGURE 5.10 Distribution of shear force along the line $y = 0.5$ m.

Due to the symmetry in the problem, the distribution of deflection and moment along y = 0.5 m is obtained with $N_S = 36$ and $N_I = 121$ and plotted in Figs. 5.11 and 5.12. It can be seen from the two figures that the proposed MFS-based meshless method provides a very accurate approximation to the corresponding analytical solution.

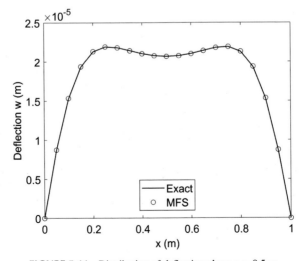

FIGURE 5.11 Distribution of deflection along y = 0.5 m.

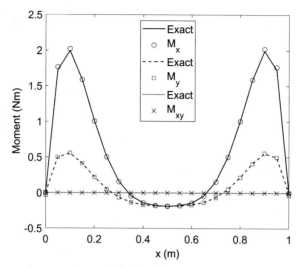

FIGURE 5.12 Distribution of moments along $y = 0.5$ m.

5.5 Remarks

In this chapter, the meshless method based on the combination of the analog equation method (AEM), RBF interpolation, and the MFS are developed for solving the thin plate bending problems with or without elastic foundation. In the light of the feature of a thin plate's governing equation, that is, fourth-order partial differential equation, the AEM is first used to convert the original governing equation into an equivalent biharmonic equation with generalized fictitious transverse load, and then the corresponding particular and homogeneous solutions are constructed by virtue of the RBF interpolation and the MFS approximation based on the improved fundamental solution, respectively. Finally, the complete satisfaction of full solution to the specific boundary conditions and the practical governing equation can determine all unknown coefficients. Numerical results show good accuracy and convergence of the proposed method.

Besides, from the generalized solution procedure in Section 5.3, the presented method is more flexible to treat various transverse load and boundary conditions, and it can be easily applied to other thin plate problems, such as thin plate with variable thickness and thin plate with heterogeneous material definitions, by just making minor changes in the final solving system of linear Eq. (5.39).

References

[1] L.H. Donnell, Beams, Plates, and Shells, McGraw-Hill, New York, 1976.
[2] S. Timoshenko, S. Woinowsky-Krieger, Theory of Plates and Shells, McGraw-Hill, New York, 1970.

[3] E. Ventsel, Thin Plates and Shells Theory: Analysis, and Applications, CRC Press, Boca Raton, 2001.

[4] J.J. Allen, Micro Electro Mechanical System Design, CRC Press, New York, 2005.

[5] A.P.S. Selvadurai, Partial Differential Equations in Mechanics, Springer, 2000.

[6] J.T. Katsikadelis, The Boundary Element Method for Engineers and Scientists: Theory and Applications, Elsevier, 2016.

[7] C.A. Brebbia, J. Dominguez, Boundary Elements: An Introductory Course, Computational Mechanics Publications, Southampton, 1992.

[8] Q.H. Qin, The Trefftz Finite and Boundary Element Method, WIT Press, Southampton, 2000.

[9] C. Cao, Q.H. Qin, A. Yu, Hybrid fundamental-solution-based FEM for piezoelectric materials, Computational Mechanics 50 (2012) 397−412.

[10] K.J. Bathe, Finite Element Procedures, Prentice-Hall, Inc., New Jersey, 1996.

[11] B. Lei, C.M. Fan, M. Li, The method of fundamental solutions for solving non-linear Berger equation of thin elastic plate, Engineering Analysis with Boundary Elements 90 (2018) 100−106.

[12] V.M.A. Leitao, A meshless method for Kirchhoff plate bending problems, International Journal for Numerical Methods in Engineering 52 (2001) 1107−1130.

[13] Q. Li, J. Soric, T. Jarak, S.N. Atluri, A locking-free meshless local Petrov-Galerkin formulation for thick and thin plates, Journal of Computational Physics 208 (2005) 116−133.

[14] H. Oh, C. Davis, J.W. Jeong, Meshfree particle methods for thin plates, Computer Methods in Applied Mechanics and Engineering (2012) 156−171.

[15] J. Sladek, V. Sladek, H.A. Mang, Meshless local boundary integral equation method for simply supported and clamped plates resting on elastic foundation, Computer Methods in Applied Mechanics and Engineering 191 (2002) 5943−5959.

[16] M. Somireddy, A. Rajagopal, Meshless natural neighbor Galerkin method for the bending and vibration analysis of composite plates, Composite Structures 111 (2014) 138−146.

[17] T. Myint-U, L. Debnath, Linear Partial Differential Equations for Scientists and Engineers, Birkhauser, Boston, 2007.

Chapter 6

Meshless analysis for two-dimensional elastic problems

Chapter outline

Methods of Fundamental Solutions in Solid Mechanics. https://doi.org/10.1016/B978-0-12-818283-3.00006-3

6.1 Introduction

As the most popular engineering problems of structural mechanics, the solutions for two-dimensional linear elasticity in isotropic solids are always of interest. For such purposes, numerical approaches have to be resorted to because of their powerful ability to deal with arbitrarily complex engineering problems. In this chapter, meshless collocation method coupling the method of fundamental solutions (MFS) and the radial basis function (RBF) interpolation is developed for the approximated solutions of two-dimensional linear elastic problems in isotropic solids with or without body forces.

To demonstrate the feasibility and ability of the present meshless method, the linear elastic problem described in Section 2.4 is extended to include the effect of thermoelasticity caused by temperature change.

As we know that when a temperature increase is added to an unconstrained elastic solid, it expands, while it shrinks when a temperature decrease happens. For example, the isotropic and homogeneous beam displayed in Fig. 6.1 will

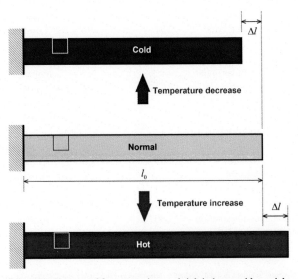

FIGURE 6.1 Schematic diagram of free expansion or shrink in beam without right end constraint.

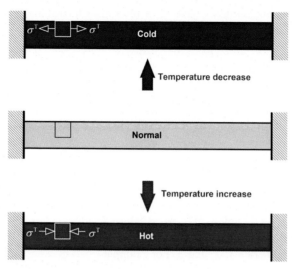

FIGURE 6.2 Schematic diagram of thermal stress and strain in beam with right end constraint.

become longer (and a little bit thinner due to Poisson's effect) when it is heated up, while it will become shorter (and a little bit wider due to Poisson's effect) when it is cooled down. Correspondingly, the relative length change caused by the temperature change in it can be written as

$$\Delta l = \alpha T l_0 \tag{6.1}$$

where Δl is the length change, l_0 is the original length, α is the thermal expansion coefficient (1/K or 1/Co), and T is the temperature change (K or Co) in the entire domain. Note that there is still no stress in the unconstrained beam in Fig. 6.1 because it will have the new length it likes to have at the given temperature.

However, if the heated region is constrained by rigid surroundings, i.e., the beam as displayed in Fig. 6.2 with constrained right end, it cannot expand freely, so it becomes subjected to compressive stress. Conversely, the cold region constrained by rigid surroundings is under tension because it cannot shrink freely. Therefore, special attention must be paid to the additional strain and stress caused by the temperature change in the solid, which are usually called as thermal strain and thermal stress, respectively. For the case of three-dimensional isotropic elastic solids under constraints, the thermal strain and stress caused by temperature change are written in tensor notation [1]:

$$\varepsilon_{ij}^{\mathrm{T}} = \alpha T \delta_{ij}, \quad i,j = 1,2,3 \tag{6.2}$$

$$\sigma_{ij}^{\mathrm{T}} = m T \delta_{ij}, \quad i,j = 1,2,3 \tag{6.3}$$

where $\varepsilon_{ij}^{\mathrm{T}}$ is the thermal strain component, σ_{ij}^{T} is the thermal stress component, δ_{ij} is the standard Delta function, and $m = \alpha E/(1 - 2\nu)$.

Hence, the constitutive equation in the three-dimensional isotropic elasticity can be modified by introducing the effect of thermal stresses as

$$\sigma_{ij} = \lambda \delta_{ij} \varepsilon_{kk} + 2\mu \varepsilon_{ij} - m\delta_{ij}T, \quad i,j = 1,2,3 \tag{6.4}$$

where

$$\lambda = \frac{2\nu}{1 - 2\nu}G, \quad \mu = G = \frac{E}{2(1 + \nu)} \tag{6.5}$$

are Lamé constants of isotropic material, and

$$\varepsilon_{kk} = \varepsilon_{11} + \varepsilon_{22} + \varepsilon_{33} = u_{k,k} \quad (k = 1,2,3) \tag{6.6}$$

represents the first invariant of strain tensor.

To make the solution procedure more compact, the tensor notation is introduced. The notation for the Cartesian components of coordinate, displacement, strain, and stress is condensed by first relabeling the x, y, and z axes by the numerals 1, 2, and 3. For example,

$$\begin{aligned}
&x_1 = x, x_2 = y, x_3 = z \\
&u_1 = u_x, u_2 = u_y, u_3 = u_z \\
&\sigma_{11} = \sigma_x, \sigma_{22} = \sigma_y, \sigma_{33} = \sigma_z, \sigma_{12} = \tau_{xy}, \sigma_{13} = \tau_{xz}, \sigma_{23} = \tau_{yz} \\
&\varepsilon_{11} = \varepsilon_x, \varepsilon_{22} = \varepsilon_y, \varepsilon_{33} = \varepsilon_z, \varepsilon_{12} = \varepsilon_{xy}, \varepsilon_{13} = \varepsilon_{xz}, \varepsilon_{23} = \varepsilon_{yz}
\end{aligned} \tag{6.7}$$

Following the reduction principle of three-dimensional isotropic elasticity to two-dimensional isotropic elasticity, the generalized governing equations of two-dimensional linear elasticity in the presence of body forces and temperature change in the elastic domain can be rewritten in tensor form as follows [1,2]:

$$\sigma_{ij,j} + b_i = 0, \quad i,j = 1,2 \tag{6.8}$$

$$\sigma_{ij} = \widetilde{\lambda} \delta_{ij} \varepsilon_{kk} + 2\widetilde{\mu} \varepsilon_{ij} - \widetilde{m}\delta_{ij}T, \quad i,j,k = 1,2 \tag{6.9}$$

$$\varepsilon_{ij} = \frac{1}{2}\left(u_{i,j} + u_{j,i}\right), \quad i,j = 1,2 \tag{6.10}$$

where b_i are body force components, and $\widetilde{\lambda}$ and $\widetilde{\mu}$ are modified materials constants

$$\widetilde{\lambda} = \frac{2\widetilde{\nu}}{1 - 2\widetilde{\nu}}G, \quad \widetilde{\mu} = G = \frac{E}{2(1 + \nu)}, \quad \widetilde{m} = \frac{\widetilde{\alpha}\widetilde{E}}{1 - 2\widetilde{\nu}} \tag{6.11}$$

with

$$\begin{cases}
\widetilde{E} = E, \quad \widetilde{\nu} = \nu, \quad \widetilde{\alpha} = \alpha & \text{for plane strain} \\
\widetilde{E} = \dfrac{1 + 2\nu}{(1 + \nu)^2}E, \quad \widetilde{\nu} = \dfrac{\nu}{1 + \nu}, \quad \widetilde{\alpha} = \dfrac{1 + \nu}{1 + 2\nu}\alpha & \text{for plane stress}
\end{cases} \tag{6.12}$$

In practice, the shear strain is the half of the engineering shear strain, that is,

$$\varepsilon_{12} = \frac{1}{2}\gamma_{12} \tag{6.13}$$

Also, the repeated subscript in Eq. (6.9) means the summation, i.e.,

$$\varepsilon_{kk} = \varepsilon_{11} + \varepsilon_{22} = u_{k,k}, \quad k = 1, 2 \tag{6.14}$$

As well, in Eqs. (6.8) and (6.10), the tensor notation is employed to represent the spatial differentiation of variables, i.e.,

$$\sigma_{ij,j} = \frac{\partial \sigma_{ij}}{\partial x_j} \quad \text{and} \quad u_{i,j} = \frac{\partial u_i}{\partial x_j} \tag{6.15}$$

Specially, if there is no temperature change in the elastic domain, the stress-strain relationship in tensor form can be obtained from Eq. (6.9) as

$$\sigma_{ij} = \widetilde{\lambda}\delta_{ij}\varepsilon_{kk} + 2\widetilde{\mu}\varepsilon_{ij}, \quad i,j,k = 1, 2 \tag{6.16}$$

Substituting Eq. (6.10) into Eq. (6.9) yields

$$\begin{aligned}
\sigma_{ij} &= \widetilde{\lambda}\delta_{ij}\varepsilon_{kk} + 2\widetilde{\mu}\varepsilon_{ij} - \widetilde{m}\delta_{ij}T \\
&= \widetilde{\lambda}\delta_{ij}u_{k,k} + \widetilde{\mu}\left(u_{i,j} + u_{j,i}\right) - \widetilde{m}\delta_{ij}T, \quad i,j,k = 1, 2
\end{aligned} \tag{6.17}$$

which are then substituted into Eq. (6.8) to give

$$\begin{aligned}
\widetilde{\lambda}\delta_{ij}u_{k,kj} &+ \widetilde{\mu}\left(u_{i,jj} + u_{j,ij}\right) - \widetilde{m}\delta_{ij}T_{,j} + b_i \\
&= \widetilde{\lambda}u_{k,ki} + \widetilde{\mu}\left(u_{i,jj} + u_{j,ij}\right) - \widetilde{m}T_{,i} + b_i \\
&= \widetilde{\lambda}u_{k,ki} + \widetilde{\mu}\left(u_{i,kk} + u_{k,ki}\right) - \widetilde{m}T_{,i} + b_i \\
&= \left(\widetilde{\lambda} + \widetilde{\mu}\right)u_{k,ki} + \widetilde{\mu}u_{i,kk} + \widetilde{b}_i = 0, \quad i,k = 1, 2
\end{aligned} \tag{6.18}$$

where

$$\widetilde{b}_i = b_i - \widetilde{m}T_{,i} \tag{6.19}$$

represent the generalized body force term.

In addition, the boundary conditions described in Eqs. (2.34) and (2.35) can be rewritten in the tensor form as

$$\begin{aligned}
u_i &= \overline{u}_i, & (i = 1, 2) & \quad on\Gamma_u \\
t_i &= \sigma_{ij}n_j = \overline{t}_i, & (i,j = 1, 2) & \quad on\Gamma_t
\end{aligned} \tag{6.20}$$

6.2 Fundamental solutions for two-dimensional elasticity

The fundamental solutions of a problem play an important role in deriving the present meshless approach, so they are first reviewed in this section.

For an infinitely extended two-dimensional linear homogeneous isotropic elastic body, as indicated in Fig. 6.3, when a unit point force is applied at

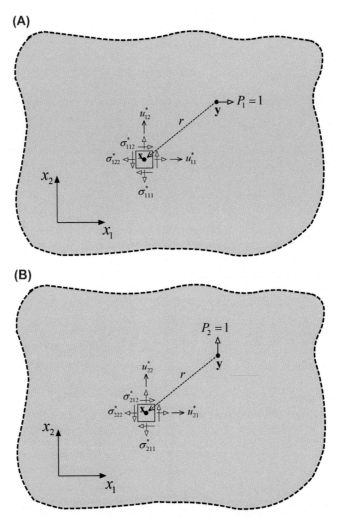

FIGURE 6.3 Definition of displacement fundamental solutions caused by unit point force (A) $P_1 = 1, P_2 = 0$ and (B) $P_1 = 0, P_2 = 1$ in an infinite two-dimensional linear homogeneous isotropic elastic body.

arbitrary location **y** in it, the induced displacement and stress solutions at arbitrary point **x** are called the fundamental solutions of the problem. To state the fundamental solutions of such a case, we shall start with the governing equation of elastostatics Eq. (6.18) in the absence of body forces and temperature change, which can be rewritten as

$$\left(\tilde{\lambda} + \tilde{\mu}\right)u_{k,ki} + \tilde{\mu}u_{i,kk} = \tilde{\mu}u_{i,kk} + \frac{\tilde{\mu}}{1 - 2\tilde{\nu}}u_{k,ki} = 0 \qquad (6.21)$$

or in matrix notation

$$\widetilde{\mu}\nabla^2\mathbf{u} + \frac{\widetilde{\mu}}{1 - 2\widetilde{v}}\nabla(\nabla\cdot\mathbf{u}) = 0 \qquad (6.22)$$

which is usually called the Cauchy−Navier operator.

Then, the fundamental solutions of displacement $u_{li}^*(\mathbf{x}, \mathbf{y})$ are required to satisfy [3]

$$\left(\widetilde{\lambda} + \widetilde{\mu}\right)u_{lk,ki}^*(\mathbf{x}, \mathbf{y}) + \widetilde{\mu}u_{li,kk}^*(\mathbf{x}, \mathbf{y}) = -\delta(\mathbf{x}, \mathbf{y})e_{li}, \quad i, k, l = 1, 2 \qquad (6.23)$$

where e_{li} is the component of the 2×2 identity matrix and $u_{li}^*(\mathbf{x}, \mathbf{y})$ represents the induced displacement in the i direction at the field point $\mathbf{x} = (x_1, x_2)$ due to the unit concentrated load acting in the l-direction at the source point $\mathbf{y} = (y_1, y_2)$.

By virtue of the complex variable method, the Lord Kelvin's method, the Fourier transform method, or the Airy stress function [2,4,5], for the plane strain state, one has

$$u_{li}^*(\mathbf{x}, \mathbf{y}) = \frac{1}{8\pi G(1 - v)}\left[(3 - 4v)\delta_{li}\ln\frac{1}{r} + r_{,l}r_{,i}\right] \qquad (6.24)$$

where $r = \|\mathbf{x} - \mathbf{y}\| = \sqrt{r_i r_i}$ is the Euclidean distance from the source point $\mathbf{y}$ subjected to the unit load at it to the field point $\mathbf{x}$ under consideration, and $r_i = x_i - y_i$ $(i = 1, 2)$.

In Eq. (6.24) and thereafter, the derivatives of the distance function r to the Cartesian coordinate component is required and can be written in tensor notations (more details can be found in Appendix A):

$$r_{,i} = \frac{\partial r}{\partial x_i} = \frac{r_i}{r}, \quad i = 1, 2 \qquad (6.25)$$

$$r_{,i}r_{,i} = \left(r_{,1}\right)^2 + \left(r_{,2}\right)^2 = \left(\frac{r_1}{r}\right)^2 + \left(\frac{r_2}{r}\right)^2 \equiv 1 \qquad (6.26)$$

Then, the first and second derivatives of the displacement fundamental solution Eq. (6.24) to the spatial variable can be given by

$$u_{li,j}^*(\mathbf{x}, \mathbf{y}) = \frac{1}{8\pi G(1 - v)}\frac{1}{r}\left[-(3 - 4v)\delta_{li}r_{,j} + r_{,i}\delta_{lj} + r_{,l}\delta_{ij} - 2r_{,i}r_{,j}r_{,l}\right] \qquad (6.27)$$

$$u_{lk,k}^*(\mathbf{x}, \mathbf{y}) = \frac{1}{4\pi G(1 - v)}\frac{1}{r}\left[-(1 - 2v)\right]r_{,l}$$

$$u_{lk,ki}^*(\mathbf{x}, \mathbf{y}) = \frac{1}{4\pi G(1 - v)}\frac{1}{r^2}(1 - 2v)\left(2r_{,l}r_{,i} - \delta_{li}\right) \qquad (6.28)$$

$$u_{li,kk}^*(\mathbf{x}, \mathbf{y}) = \frac{1}{4\pi G(1 - v)}\frac{1}{r^2}\left(\delta_{li} - 2r_{,i}r_{,l}\right)$$

Therefore, from Eq. (6.10), we can obtain the following fundamental solutions of strain:

$$
\varepsilon^*_{lij}(\mathbf{x},\mathbf{y}) = \frac{1}{2}\left[u^*_{li,j}(\mathbf{x},\mathbf{y}) + u^*_{lj,i}(\mathbf{x},\mathbf{y})\right]
$$
$$
= \frac{1}{8\pi G(1-v)}\frac{1}{r}\left[-(1-2v)\left(r_{,j}\delta_{li} + r_{,i}\delta_{lj}\right) + r_{,l}\delta_{ij} - 2r_{,i}r_{,j}r_{,l}\right] \tag{6.29}
$$

from which we have

$$
\varepsilon^*_{lkk}(\mathbf{x},\mathbf{y}) = u^*_{lk,k} = \frac{1}{4\pi G(1-v)}\frac{1}{r}\left[-(1-2v)\right]r_{,l} \tag{6.30}
$$

As a result, the fundamental solutions of stress denoted by $\sigma^*_{lij}(\mathbf{x},\mathbf{y})$ can be obtained by means of the stress-strain relationships Eq. (6.16) without the generalized body force term as

$$
\sigma^*_{lij}(\mathbf{x},\mathbf{y}) = \tilde{\lambda}\delta_{ij}\varepsilon^*_{lkk}(\mathbf{x},\mathbf{y}) + 2\tilde{\mu}\varepsilon^*_{lij}(\mathbf{x},\mathbf{y})
$$
$$
= \frac{1}{4\pi(1-v)r}\left[(1-2v)(r_{,l}\delta_{ij} - r_{,j}\delta_{il} - r_{,i}\delta_{jl}) - 2r_{,i}r_{,j}r_{,l}\right],\ i,j,l = 1,2 \tag{6.31}
$$

Finally, the corresponding fundamental solutions of boundary traction with respect to a smooth boundary section defined by the outward normal $\mathbf{n} = (n_1, n_2)$ can be derived by

$$
t^*_{li}(\mathbf{x},\mathbf{y}) = \sigma^*_{lij}(\mathbf{x},\mathbf{y})n_j(\mathbf{x})
$$
$$
= \frac{1}{4\pi(1-v)r}\left\{(1-2v)(r_{,l}n_i - r_{,i}n_l) - \frac{\partial r}{\partial n}\left[(1-2v)\delta_{il} + 2r_{,i}r_{,l}\right]\right\},\ i,j,l = 1,2 \tag{6.32}
$$

The preceding formulations are obtained for the case of plane strain, while the fundamental solutions for the plane stress case can be obtained by a simple replacement in Eqs. (6.24), (6.29), (6.31) and (6.32):

$$
E \rightarrow \frac{(1+2v)E}{(1+v)^2}, \qquad v \rightarrow \frac{v}{1+v} \tag{6.33}
$$

6.3 Solution procedure for homogeneous elasticity

6.3.1 Solution procedure

If the problem is in absence of any generalized body forces and the isotropic elastic material involved has constant elastic modulus and Poisson's ratio, as indicated in Fig. 6.4, the governing Eq. (6.18) reduces to the homogeneous one as follows [2,4]:

$$
(\tilde{\lambda} + \tilde{\mu})u_{k,ki} + \tilde{\mu}u_{i,kk} = 0, \quad i,k = 1,2 \tag{6.34}
$$

which can be solved by the standard method of fundamental solutions directly [6].

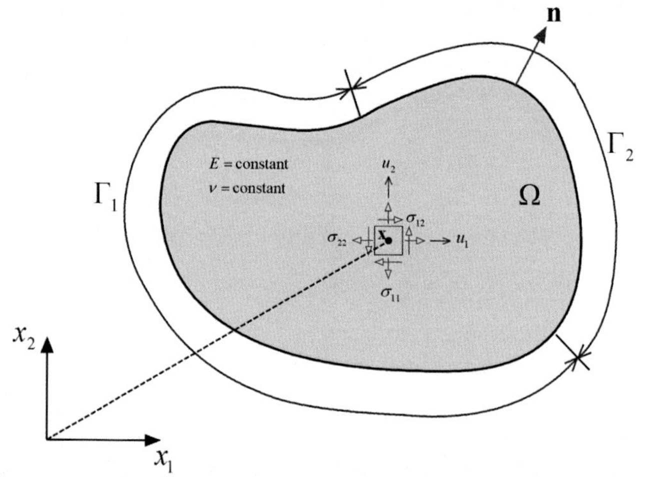

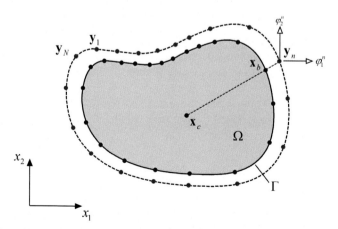

FIGURE 6.4 Two-dimensional isotropic and homogeneous elasticity.

By means of the definition of fundamental solutions of the isotropic and homogenous elastic solids, the displacement solution to Eq. (6.34) at arbitrary field point **x** in the domain Ω can be assumed to be the linear combination of fundamental solutions centered at different source points $\mathbf{y}_n$ $(n = 1, 2, ..., N)$, that is,

$$u_i(\mathbf{x}) = \sum_{n=1}^{N} \varphi_l^n u_{li}^*(\mathbf{x}, \mathbf{y}_n), \quad \forall \mathbf{x} \in \Omega, \ \mathbf{y}_n \notin \Omega \tag{6.35}$$

where N is the number of source points outside the domain and φ_l^n is the source intensity along the l-direction at the nth source point, as illustrated in Fig. 6.5. Besides, the repeated subscript $l(l = 1, 2)$ in Eq. (6.35) means the summation.

FIGURE 6.5 Boundary collocations for two-dimensional isotropic and homogeneous elasticity without body forces.

In matrix form, Eq. (6.35) can be rewritten as

$$\mathbf{u}(\mathbf{x}) = \mathbf{U}(\mathbf{x})\boldsymbol{\varphi}, \quad \forall \mathbf{x} \in \Omega, \mathbf{y}_n \notin \Omega \qquad (6.36)$$

where

$$\mathbf{u}(\mathbf{x}) = \left\{ \begin{array}{c} u_1(\mathbf{x}) \\ u_2(\mathbf{x}) \end{array} \right\} \qquad (6.37)$$

is the desired displacement vector at the field point $\mathbf{x}$,

$$\mathbf{U}(\mathbf{x}) = [\mathbf{u}^*(\mathbf{x}, \mathbf{y}_1) \quad \mathbf{u}^*(\mathbf{x}, \mathbf{y}_2) \quad \cdots \quad \mathbf{u}^*(\mathbf{x}, \mathbf{y}_N)]_{2 \times 2N} \qquad (6.38)$$

is the so-called displacement fundamental solution matrix consisting of the following 2×2 submatrices

$$\mathbf{u}^*(\mathbf{x}, \mathbf{y}_n) = \begin{bmatrix} u_{11}^*(\mathbf{x}, \mathbf{y}_n) & u_{21}^*(\mathbf{x}, \mathbf{y}_n) \\ u_{12}^*(\mathbf{x}, \mathbf{y}_n) & u_{22}^*(\mathbf{x}, \mathbf{y}_n) \end{bmatrix}_{2 \times 2}, \quad n = 1, 2, ..., N \qquad (6.39)$$

and

$$\boldsymbol{\varphi} = \left\{ \begin{array}{c} \varphi_1^1 \\ \varphi_2^1 \\ \hline \varphi_1^2 \\ \varphi_2^2 \\ \hline \vdots \\ \vdots \\ \hline \varphi_1^N \\ \varphi_2^N \end{array} \right\}_{2N \times 1} = \left\{ \begin{array}{c} \boldsymbol{\varphi}^1 \\ \boldsymbol{\varphi}^2 \\ \vdots \\ \boldsymbol{\varphi}^N \end{array} \right\}_{2N \times 1} \qquad (6.40)$$

is the vector of unknown coefficients with

$$\boldsymbol{\varphi}^n = \left\{ \begin{array}{c} \varphi_1^n \\ \varphi_2^n \end{array} \right\}_{2 \times 1}, \quad n = 1, 2, ..., N \qquad (6.41)$$

Obviously, the approximated displacement solution depicted in Eq. (6.35) exactly satisfies the homogeneous governing Eq. (6.34). Therefore, only boundary conditions should be satisfied if such displacement approximation is employed.

From Eqs. (6.10) and (6.16), the strain and stress components can be obtained using the approximated displacement field and written in the following form of collocation summation:

$$\varepsilon_{ij}(\mathbf{x}) = \sum_{n=1}^{N} \varphi_l^n \varepsilon_{lij}^*(\mathbf{x}, \mathbf{y}_n)$$

$$\sigma_{ij}(\mathbf{x}) = \sum_{n=1}^{N} \varphi_l^n \sigma_{lij}^*(\mathbf{x}, \mathbf{y}_n) \qquad (6.42)$$

which can be rewritten in the following matrix form

$$\boldsymbol{\varepsilon}(\mathbf{x}) = \mathbf{B}(\mathbf{x})\boldsymbol{\varphi}$$
$$\boldsymbol{\sigma}(\mathbf{x}) = \mathbf{S}(\mathbf{x})\boldsymbol{\varphi} \tag{6.43}$$

where

$$\boldsymbol{\varepsilon}(\mathbf{x}) = \left\{ \begin{array}{c} \varepsilon_{11}(\mathbf{x}) \\ \varepsilon_{22}(\mathbf{x}) \\ \gamma_{12}(\mathbf{x}) \end{array} \right\}, \quad \boldsymbol{\sigma}(\mathbf{x}) = \left\{ \begin{array}{c} \sigma_{11}(\mathbf{x}) \\ \sigma_{22}(\mathbf{x}) \\ \sigma_{12}(\mathbf{x}) \end{array} \right\} \tag{6.44}$$

are the strain and stress vectors, and

$$\mathbf{B}(\mathbf{x}) = [\boldsymbol{\varepsilon}^*(\mathbf{x}, \mathbf{y}_1) \quad \boldsymbol{\varepsilon}^*(\mathbf{x}, \mathbf{y}_2) \quad \cdots \quad \boldsymbol{\varepsilon}^*(\mathbf{x}, \mathbf{y}_N)]_{3 \times 2N}$$
$$\mathbf{S}(\mathbf{x}) = [\boldsymbol{\sigma}^*(\mathbf{x}, \mathbf{y}_1) \quad \boldsymbol{\sigma}^*(\mathbf{x}, \mathbf{y}_2) \quad \cdots \quad \boldsymbol{\sigma}^*(\mathbf{x}, \mathbf{y}_N)]_{3 \times 2N} \tag{6.45}$$

are the fundamental solution matrices of strain and stress consisting of 3×2 submatrices:

$$\boldsymbol{\varepsilon}^*(\mathbf{x}, \mathbf{y}_n) = \begin{bmatrix} \varepsilon^*_{111}(\mathbf{x}, \mathbf{y}_n) & \varepsilon^*_{211}(\mathbf{x}, \mathbf{y}_n) \\ \varepsilon^*_{122}(\mathbf{x}, \mathbf{y}_n) & \varepsilon^*_{222}(\mathbf{x}, \mathbf{y}_n) \\ \varepsilon^*_{112}(\mathbf{x}, \mathbf{y}_n) & \varepsilon^*_{212}(\mathbf{x}, \mathbf{y}_n) \end{bmatrix}_{3 \times 2} , \quad n = 1, 2, \ldots, N$$

$$\boldsymbol{\sigma}^*(\mathbf{x}, \mathbf{y}_n) = \begin{bmatrix} \sigma^*_{111}(\mathbf{x}, \mathbf{y}_n) & \sigma^*_{211}(\mathbf{x}, \mathbf{y}_n) \\ \sigma^*_{122}(\mathbf{x}, \mathbf{y}_n) & \sigma^*_{222}(\mathbf{x}, \mathbf{y}_n) \\ \sigma^*_{112}(\mathbf{x}, \mathbf{y}_n) & \sigma^*_{212}(\mathbf{x}, \mathbf{y}_n) \end{bmatrix}_{3 \times 2} , \quad n = 1, 2, \ldots, N \tag{6.46}$$

Further, the traction components over the boundary can be expressed as

$$t_i(\mathbf{x}) = \sum_{n=1}^{N} \varphi_l^n t_{li}^*(\mathbf{x}, \mathbf{y}_n), \quad \forall \mathbf{x} \in \Gamma, \mathbf{y}_n \notin \Omega \tag{6.47}$$

or in matrix form

$$\mathbf{t}(\mathbf{x}) = \mathbf{T}(\mathbf{x})\boldsymbol{\varphi}, \quad \forall \mathbf{x} \in \Gamma, \mathbf{y}_n \notin \Omega \tag{6.48}$$

where

$$\mathbf{t}(\mathbf{x}) = \left\{ \begin{array}{c} t_1(\mathbf{x}) \\ t_2(\mathbf{x}) \end{array} \right\} \tag{6.49}$$

is the traction vector at the field point $\mathbf{x}$ on the boundary, and

$$\mathbf{T}(\mathbf{x}) = [\mathbf{t}^*(\mathbf{x}, \mathbf{y}_1) \quad \mathbf{t}^*(\mathbf{x}, \mathbf{y}_2) \quad \cdots \quad \mathbf{t}^*(\mathbf{x}, \mathbf{y}_N)]_{2 \times 2N} \tag{6.50}$$

is the traction fundamental solution matrix consisting of the following 2×2 submatrices:

$$\mathbf{t}^*(\mathbf{x}, \mathbf{y}_n) = \begin{bmatrix} t^*_{11}(\mathbf{x}, \mathbf{y}_n) & t^*_{21}(\mathbf{x}, \mathbf{y}_n) \\ t^*_{12}(\mathbf{x}, \mathbf{y}_n) & t^*_{22}(\mathbf{x}, \mathbf{y}_n) \end{bmatrix}_{2 \times 2} , \quad n = 1, 2, \ldots, N \tag{6.51}$$

To determine the unknown coefficient vector φ, a series of boundary collocations or boundary nodes may be chosen on the physical boundary, and then we can enforce the approximations Eqs. (6.36) and (6.48) to satisfy the specific displacement and traction conditions at these boundary collocations. To do this, L boundary collocations ($L \geq N$) are chosen. At these points, we have

$$\sum_{n=1}^{N} \varphi_l^n u_{li}^*\left(\mathbf{x}_k^u, \mathbf{y}_n\right) = \bar{u}_i\left(\mathbf{x}_k^u\right), \quad k = 1 \rightarrow L_1, \quad \mathbf{x}_k^u \in \Gamma_1$$

$$\sum_{n=1}^{N} \varphi_l^n t_{li}^*\left(\mathbf{x}_k^t, \mathbf{y}_n\right) = \bar{t}_i\left(\mathbf{x}_k^t\right), \quad k = 1 \rightarrow L_2, \quad \mathbf{x}_k^t \in \Gamma_2 \tag{6.52}$$

which can be rewritten in matrix form:

$$\begin{bmatrix} \mathbf{U}\left(\mathbf{x}_1^u\right) \\ \vdots \\ \mathbf{U}\left(\mathbf{x}_{L_1}^u\right) \\ \hline \mathbf{T}\left(\mathbf{x}_1^t\right) \\ \vdots \\ \mathbf{T}\left(\mathbf{x}_{L_2}^t\right) \end{bmatrix}_{2L \times 2N} \boldsymbol{\varphi} = \begin{bmatrix} \bar{\mathbf{u}}\left(\mathbf{x}_1^u\right) \\ \vdots \\ \bar{\mathbf{u}}\left(\mathbf{x}_{L_1}^u\right) \\ \hline \bar{\mathbf{t}}\left(\mathbf{x}_1^t\right) \\ \vdots \\ \bar{\mathbf{t}}\left(\mathbf{x}_{L_2}^t\right) \end{bmatrix}_{2L \times 1} \tag{6.53}$$

where L_1 and L_2 are the number of boundary collocations on the displacement and traction boundaries Γ_1 and Γ_2, respectively, and $L_1 + L_2 = L$.

Explicitly, this leads to a system consisting of $2L$ linear equations with $2N$ unknown coefficients:

$$\mathbf{H}_{2L \times 2N} \boldsymbol{\varphi}_{2N \times 1} = \mathbf{f}_{2L \times 1} \tag{6.54}$$

where $\mathbf{H}$ is the $2L \times 2N$ collocation matrix with elements H_{ij}, $\boldsymbol{\varphi}$ is the $2N \times 1$ vector of unknowns, and $\mathbf{f}$ is the $2L \times 1$ known vector.

Eq. (6.53) or Eq. (6.54) is the final set of discrete equations for the implementation of the MFS. Solving Eq. (6.53) by the standard Gaussian Elimination if $L = N$ or the least square method if $L > N$, the unknown coefficients can be determined numerically. The default setting in the practical computation is $L = N$. After obtaining the interpolating coefficients, the displacements and stresses at an arbitrary point in the domain can be naturally evaluated using Eqs. (6.35) and (6.42).

6.3.2 Program structure and source code

Based on the MFS formulation described before, MATLAB codes can be programmed for the numerical analysis of two-dimensional elasticity problems. The main solution procedures of the program include the following:

- Read input data and allocate proper array sizes.
- **For** each boundary collocation **do** the following:
  - Compute coefficients H_{ij}.
  - Assemble to form coefficient matrix $\mathbf{H}$.

- Solve the resulting matrix equation for the approximating coefficients.
- Compute the displacement and stress results at test points.
- Output the displacement and stress solutions.

6.3.2.1 Input data

The major input parameters needed for the implementation of MFS are these:

PT: type of plane elastic problem (=1: plane strain; =2: plane stress)
ND: number of dimensions of problem
NF: number of degrees of freedom (DOFs) per point
NR: number of boundary collocations on the real boundary
NV: number of virtual source points
NT: number of test points in the domain
RC: NR by ND matrix storing coordinates of each boundary collocation
RN: NR by ND matrix storing normal components at each boundary collocation
BT: NR by NF matrix storing type of boundary conditions at each boundary collocation
BV: NR by NF matrix storing given value of boundary conditions at each boundary collocation
VC: NV by ND matrix storing coordinates of virtual source points
TC: NT by ND matrix storing coordinates of test points

Most of these parameters input from data files are associated with the geometric information about the solution domain and its given boundary conditions of a problem defined by a user. The boundary collocations can be generated either manually or using an automatic boundary discretization program, and they can then be used to generate source points outside the domain using Eq. (1.48). Additionally, the information for the given boundary conditions at each boundary collocation includes the type of boundary condition and the given constraint value along the given DOF. For the present two-dimensional elasticity problem, the number of degrees of freedom of each boundary collocation is $NF = 2$, which means that the displacement or traction force component should be given along each DOF at the boundary collocation, as indicated in Fig. 6.6. For example, at any constrained boundary collocation i ($i = 1, 2, \ldots , NR$), the given boundary condition information can be written as

6.3.2.1.1 First DOF

$BT[i, 1] = 0$: specified displacement u_1
$BT[i, 1] = 1$: specified traction t_1
$BV[i, 1] =$ specified value along the first DOF

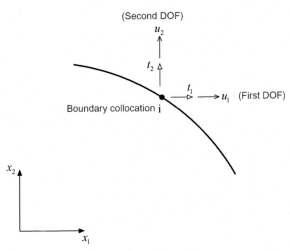

FIGURE 6.6 Boundary condition information at boundary collocation for two-dimensional elasticity.

6.3.2.1.2 Second DOF

BT[i, 2] = 0: specified displacement u_2
BT[i, 2] = 1: specified traction t_2
BV[i, 2] = specified value along the second DOF

6.3.2.2 Computation of coefficient matrix

The resulting coefficient matrix **H** is expressed in Eq. (6.54). The size of the **H** matrix is determined by the number of boundary collocations and source points prepared in the input part, and each element in it is computed from the fundamental solution u_{li}^* $(l, i = 1, 2)$ or t_{li}^* $(l, i = 1, 2)$ given in Eqs. (6.24) and (6.32) with the coordinates of the specific boundary collocation and source point.

6.3.2.3 Solving the resulting system of linear equations

Once the system of linear equations is established, as indicated in Eq. (6.54), it can be solved for the approximating coefficients φ^n $(n = 1, 2, \ldots, N)$ at N boundary collocations, which are the primary unknowns in the MFS. In the MATLAB program, such systems can be solved by the inbuilt function, that is, the matrix left division:

$$\varphi = \mathbf{H} \backslash \mathbf{f}$$

Subsequently, the physical quantities at any point in the domain can be evaluated with the solved coefficients φ^n.

6.3.2.4 Source code

Source code is written in M-function in MATLAB, and the main function MFS_2DElasticity.m calls another M-function (subroutine) FDS2DET.m that is coded to evaluate the fundamental solutions of displacement u_{li}^* and stress σ_{lij}^* when the coordinates of source point and field point are input. The flowchart of the program is given in Fig. 6.7.

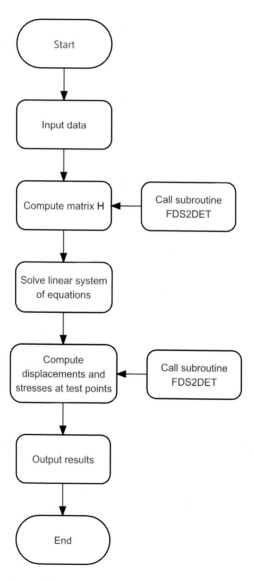

FIGURE 6.7 Flowchart of the program of MFS_2DElasticity.m for solving two-dimensional elasticity problems.

```
% ======================================================
% Main program to solve 2D Elasticity problem using the MFS
% ======================================================
function MFS_2DElasticity
%
% **** Variable statements
% PT: Type of plane elastic problem (=1: plane strain; =2: plane
%    stress)
% ND: Number of dimensions of problem
% NF: Number of DOFs of problem at every point
% NR: Number of boundary collocations on the real boundary
% NV: Number of virtual source points
% NT: Number of test points in the domain
%
% RC: NR by ND matrix of coordinates of boundary collocations
% RN: NR by ND matrix of normals at boundary collocations
% BT: NR by NF matrix of types of boundary conditions at boundary
%    collocations
% BV: NR by NF matrix of values of boundary conditions
% VC: NV by ND matrix of coordinates of virtual source points
% TC: NT by ND matrix of coordinates of test points
%
% E, MU: material properties
%---------------------------------   ----------------------------
clear all;
close all;

disp('**** Select type of plane elastic problems :');
disp('    1) Plane Strain Analysis');
disp('    2) Plane Stress Analysis');
disp('-----------------------------');
PT=input('Choose 1 or 2 and then press enter :');

% ** Open data file for input
fp=fopen('Input.txt','rt');
dummy=char(zeros(1,100));
% Test description
dummy=fgets(fp);
% Basic parameters
dummy=fgets(fp);
TMP=str2num(fgets(fp));
[ND,NF,NR,NV,NT]=deal(TMP(1),TMP(2),TMP(3),TMP(4),TMP(5));
% Initialization
RC=zeros(NR,ND);
RN=zeros(NR,ND);
BT=zeros(NR,NF);
BV=zeros(NR,NF);
```

```
VC=zeros(NV,ND);
TC=zeros(NT,ND);
% Boundary collocation coordinates and normals
dummy=fgets(fp);
dummy=fgets(fp);
for i=1:NR
    TMP=str2num(fgets(fp));
    [NUM,RC(i,1:ND),RN(i,1:ND)]=deal(TMP(1),TMP(2:1+ND),
    TMP(2+ND:1+2*ND));
end
% Type and value of boundary conditions
dummy=fgets(fp);
dummy=fgets(fp);
for i=1:NR
    TMP=str2num(fgets(fp));
    [NUM,BT(i,1:NF),BV(i,1:NF)]=deal(TMP(1),TMP(2:1+NF),
    TMP(2+NF:1+2*NF));
end
% Source points
dummy=fgets(fp);
dummy=fgets(fp);
for i=1:NV
    TMP=str2num(fgets(fp));
    [NUM,VC(i,1:ND)]=deal(TMP(1),TMP(2:1+ND));
end
% Test points
dummy=fgets(fp);
dummy=fgets(fp);
for i=1:NT
    TMP=str2num(fgets(fp));
    [NUM,TC(i,1:ND)]=deal(TMP(1),TMP(2:1+ND));
End
% Material properties
dummy=fgets(fp);
dummy=fgets(fp);
TMP=str2num(fgets(fp));
[E,MU]=deal(TMP(1),TMP(2));
fclose(fp);
% Form the coefficient matrix and right-handed term
nrow=NR*NF;
ncol=NV*NF;
HH=zeros(nrow, ncol);
FF=zeros(nrow,1);
for i=1:NR
    x=RC(i,1);
    y=RC(i,2);
```

```
    nx=RN(i,1);
    ny=RN(i,2);
    irow=i*NF;
    for j=1:NV
      vx=VC(j,1);
      vy=VC(j,2);
      jcol=j*NF;
      [h11,h12,h22,g111,g112,g122,g211,g212,g222]=
      FDS2DET(x,y,vx,vy,ME,MMU);
      if BT(i,1)==0 % Specified displacement in the x direction
        HH(irow-1,jcol-1)=h11;
        HH(irow-1,jcol) =h12;
      end
      if BT(i,1)==1 % Specified traction in the x direction
        HH(irow-1,jcol-1)=g111*nx+g112*ny;
        HH(irow-1,jcol) =g211*nx+g212*ny;
      end
      if BT(i,2)==0 % Specified displacement in the y direction
        HH(irow,jcol-1)=h12;
        HH(irow, jcol) =h22;
      end
      if BT(i,2)==1 % Specified traction in the y direction
        HH(irow,jcol-1)=g112*nx+g122*ny;
        HH(irow, jcol) =g212*nx+g222*ny;
      end
    end
    FF(irow-1,1)=BV(i,1);
    FF(irow,1) =BV(i,2);
end
%------------------------------------------------------------------
% ** Solve the linear system of equations
FF=HH\FF; % Solve Ax=b using matrix left division
%------------------------------------------------------------------
% **Evaluate quantities at test points
UT=zeros(NT,2);
ST=zeros(NT,3);
for i=1:NT
  x=TC(i,1);
  y=TC(i,2);
  for j=1:NNV
    vx=VC(j,1);
    vy=VC(j,2);
    jcol=j*NDN;
    [h11,h12,h22,g111,g112,g122,g211,g212,g222]=
    FDS2DET(x,y,vx,vy,ME,MMU);
```

```
        UT(i,1)=UT(i,1)+FF(jcol-1)*h11+FF(jcol)*h12; % U1
        UT(i,2)=UT(i,2)+FF(jcol-1)*h12+FF(jcol)*h22; % U2
        ST(i,1)=ST(i,1)+FF(jcol-1)*g111+FF(jcol)*g211; % S11
        ST(i,2)=ST(i,2)+FF(jcol-1)*g122+FF(jcol)*g222; % S22
        ST(i,3)=ST(i,3)+FF(jcol-1)*g112+FF(jcol)*g212; % S12
    end
end
%------------------------------------------------------------
%**Output results in Cartesian coordinate system at test points
fp=fopen('Results.txt','wt');
fprintf(fp,'No. x, y, U1, U2, S11, S22, S12\n');
for i=1:NT
    x=TC(i,1);
    y=TC(i,2);
    u1=UT(i,1);
    u2=UT(i,2);
    s11=ST(i,1);
    s22=ST(i,2);
    s12=ST(i,3);
fprintf(fp,'%3d, %5.3e, %5.3e, %9.6e, %9.6e, %9.6e, %9.6e, %9.6e\n',
i,x,y,u1,u2,s11,s22,s12);
end
fclose(fp);
close all;
%-------------End mian program--------------------
% ====================================================
% Subroutine FDS2DET: Compute the 2D fundamental solutions at the field
point
% (x,y) with the given source point (vx,vy)
% ====================================================
function
[h11,h12,h22,g111,g112,g122,g211,g212,g222]=
FDS2DET(x,y,vx,vy,ME,MMU)
%
rx=x-vx;
ry=y-vy;
r=sqrt(rx^2+ry^2);
rx=rx/r; % dr/dx
ry=ry/r; % dr/dy
% Displacement parts
t=ME/(1+MMU)/2;
t1=1/8/pi/t/(1-MMU);
t2=3-4*MMU;
h11=t1*(t2*log(1/r)+rx*rx);
h12=t1*rx*ry;
h22=t1*(t2*log(1/r)+ry*ry);
```

```
% Stress parts
t1=1/4/pi/(1-MMU);
t2=1-2*MMU;
g111=t1/r*(t2*(-rx)-2*rx^3);
g112=t1/r*(t2*(-ry)-2*rx^2*ry);
g122=t1/r*(t2*rx-2*rx*ry^2);

g211=t1/r*(t2*ry-2*rx^2*ry);
g212=t1/r*(t2*(-rx)-2*rx*ry^2);
g222=t1/r*(t2*(-ry)-2*ry^3);
```

6.3.3 Results and discussion

In investigating the performance of MFS for homogeneous elastic problems, two examples are considered, including a thick-walled cylinder under internal pressure and an infinite plate with centered circular hole subjected to a remote tension. Moreover, the absolute error (*Aerr*) and relative error (*Rerr*) of variable ζ is defined as for analysis of accuracy and convergence of numerical solutions:

$$Aerr(\zeta) = |\zeta - \tilde{\zeta}| \tag{6.55}$$

$$Rerr(\zeta) = \frac{|\zeta - \tilde{\zeta}|}{|\zeta|} \tag{6.56}$$

where ζ and $\tilde{\zeta}$ are exact and numerical solutions at a specific point, respectively.

6.3.3.1 Thick-walled cylinder under internal pressure

Consider an infinitely long thick-walled circular cylinder subjected to internal pressure, as displayed in Fig. 6.8. The problem is a very common one in pressure vessels such as boiler drums and chemical reaction vessels. In computation, the inner and outer radiuses are 50 and 100 mm, respectively. The internal pressure is 0.4 MPa. The Young's modulus $E = 2.1 \times 10^4$ MPa

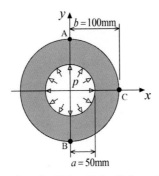

FIGURE 6.8 Illustration of a thick-walled cylinder under internal pressure.

and the Poisson's ratio $v = 0.3$. The exact solutions in polar coordinates for plane stress case are given as

$$u_r = \frac{1}{E}\left[-(1+v)\frac{A}{r} + 2(1-v)rC\right]$$

$$u_\theta = 0$$

$$\sigma_r = \frac{A}{r^2} + 2C \qquad\qquad (6.57)$$

$$\sigma_\theta = -\frac{A}{r^2} + 2C$$

$$\tau_{r\theta} = 0$$

where

$$A = -\frac{a^2 b^2 p_0}{b^2 - a^2}, \quad C = \frac{a^2 p_0}{2(b^2 - a^2)} \qquad\qquad (6.58)$$

For the plane strain case under consideration, we just need to replace E and v with $\frac{E}{1-v^2}$ and $\frac{v}{1-v}$ respectively in Eq. (6.57) to give the related displacement and stress solutions.

To avoid rigid body motion, some displacement components are constrained. In this example, they are the displacement components in the x direction at points A and B and the displacement component in y direction at point C (see Fig. 6.8).

For this double domain, the virtual boundary the source points locate consists of the outer boundary and the inner boundary, as depicted in Fig. 6.9. The corresponding similarity ratio between the virtual boundary and the real

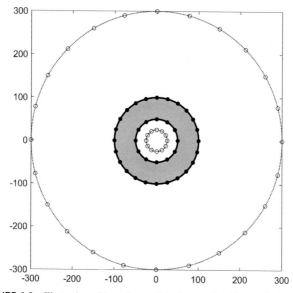

FIGURE 6.9 Illustration of boundary collocations in the thick-walled cylinder.

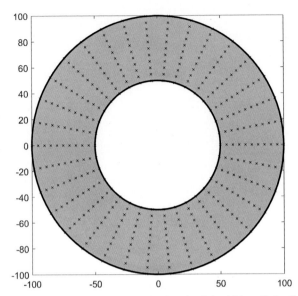

FIGURE 6.10 Computing points in the domain of the thick-walled cylinder.

boundary is 3 for the outer virtual boundary and 0.6 for the inner virtual boundary. Simultaneously, some computing points are chosen in the cylinder, so numerical results at these points can be evaluated and recorded for further treatment, as indicated in Fig. 6.10. Figs. 6.11−6.13 reveal the variation of absolute error in the field variables u_r, σ_r, and σ_θ between the numerical results and the analytical solutions, from which it is observed that as the number of boundary collocations increases, the results improve. This can be regarded as demonstration of the convergence of the MFS. As well, in Figs. 6.14 and 6.15,

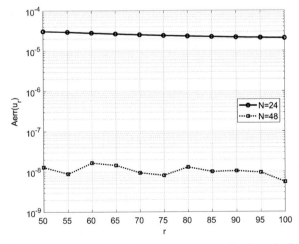

FIGURE 6.11 Absolute error in radial displacement in terms of radius r in the thick-walled cylinder.

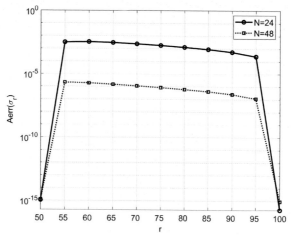

FIGURE 6.12 Absolute error in radial stress in terms of radius r in the thick-walled cylinder.

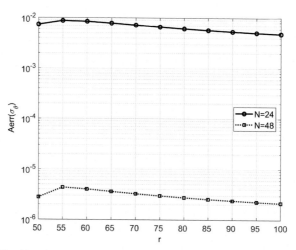

FIGURE 6.13 Absolute error in hoop stress in terms of radius r in the thick-walled cylinder.

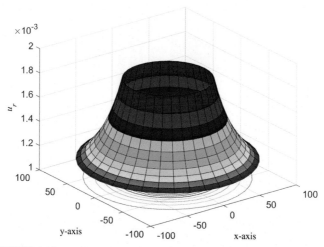

FIGURE 6.14 Variations of radial displacement in the thick-walled cylinder.

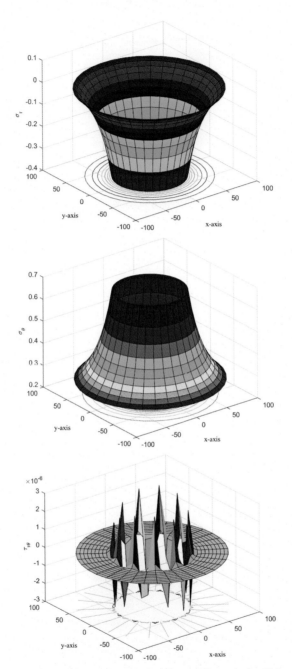

FIGURE 6.15 Variations of stress components in polar coordinates in the thick-walled cylinder.

the distributions of radial displacement u_r and the normal stress components σ_r and σ_θ in the entire domain are plotted. It is evident that these field variables with respect to the polar coordinates show apparent dependence of the radial coordinate r. From Fig. 6.15, however, the variation of shear stress $\tau_{r\theta}$ in the domain looks irregular, due to the numerical float errors, but its magnitude is very small (about 10^{-6}), compared to that of normal stresses σ_r and σ_θ. Finally, the variations of displacements and stresses in terms of Cartesian coordinates in the thick-wall cylinder domain are plotted in Figs. 6.16 and 6.17 for comparison. We observe that all displacement components and stress components in Cartesian coordinates are not zero, although $u_\theta = 0$ and $\tau_{r\theta} = 0$.

Of special note, $A \rightarrow -a^2 p_0$ and $C \rightarrow 0$ when $b \rightarrow \infty$. For this special case, the problem can be converted into an infinite domain with circular hole under internal pressure, as shown in Fig. 6.18. The corresponding solutions can be written as

$$u_r = \frac{1}{E}\left[-(1+\nu)\frac{A}{r}\right] = -\frac{1}{2G}\frac{A}{r}, \quad u_\theta = 0$$

$$\sigma_r = \frac{A}{r^2}, \qquad \sigma_\theta = -\frac{A}{r^2}, \qquad \tau_{r\theta} = 0 \tag{6.59}$$

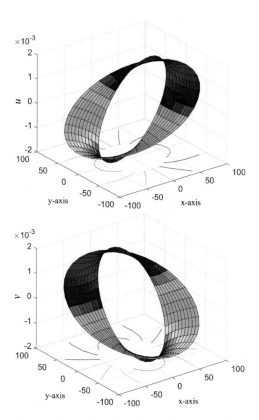

FIGURE 6.16 Variations of displacement components in Cartesian coordinates in the thick-walled cylinder.

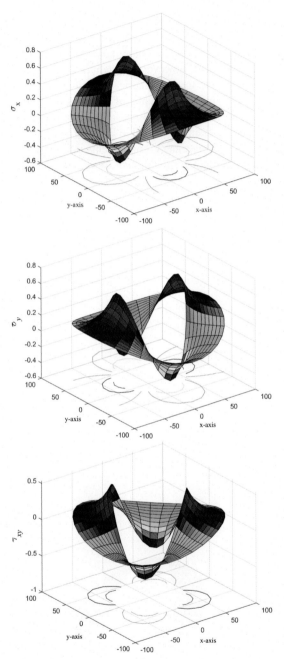

FIGURE 6.17 Variations of stress components in Cartesian coordinates in the thick-walled cylinder.

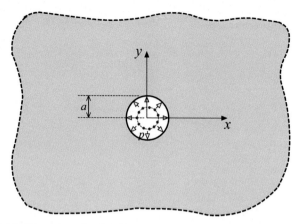

FIGURE 6.18 Illustration of an infinite domain with circular hole under internal pressure.

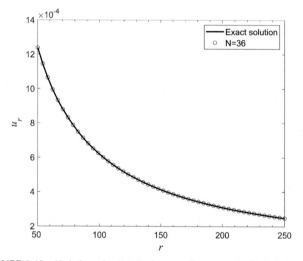

FIGURE 6.19 Variation of radial displacement in terms of r for infinite case.

Figs. 6.19 and 6.20 show the variations of u_r and σ_r, σ_θ in terms of the radial coordinate r when the number of boundary collocations is 36, and it is clearly shown that the numerical results agree well with the analytical solutions evaluated by Eq. (6.59).

6.3.3.2 Infinite domain with circular hole subjected to a far-field remote tensile

In the example shown in Fig. 6.21, an infinite domain with a circular hole subjected to an infinitely remote uniformly distributed tensile stress is considered.

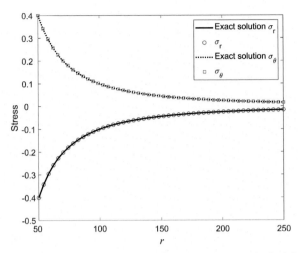

FIGURE 6.20   Variation of radial stress and hoop stress in terms of r for infinite case.

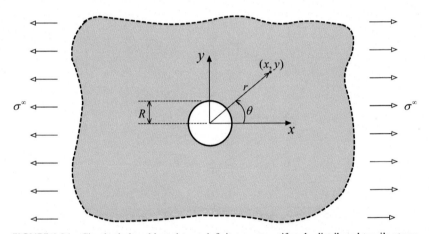

FIGURE 6.21   Circular hole subjected to an infinite remote uniformly distributed tensile stress.

In the computation, $R = 0.4$ mm. The far-field stress $\sigma^\infty = 1$ MPa, the Young's modulus $E = 2.1 \times 10^4$ MPa, and Poisson ratio $\nu = 0.3$. The exact stress solutions of the original problem in polar coordinates are given as

$$\sigma_r = \frac{\sigma^\infty}{2}\left(1 - \frac{R^2}{r^2}\right) + \frac{\sigma^\infty}{2}\left(1 + \frac{3R^4}{r^4} - 4\frac{R^2}{r^2}\right)\cos(2\theta)$$

$$\sigma_\theta = \frac{\sigma^\infty}{2}\left(1 + \frac{R^2}{r^2}\right) - \frac{\sigma^\infty}{2}\left(1 + \frac{3R^4}{r^4}\right)\cos(2\theta) \qquad (6.60)$$

$$\sigma_{r\theta} = -\frac{\sigma^\infty}{2}\left(1 - \frac{3R^4}{r^4} + 2\frac{R^2}{r^2}\right)\sin(2\theta)$$

From the mechanical point of view, the original problem can be viewed as the superposition of two subproblems. One is an infinite domain under remote tension, and the other is an infinite domain with circular hole subjected to traction along x direction on the circular boundary. Clearly, the stress solution for the former subproblem is constant:

$$\sigma_x^{(1)} = 0, \sigma_{xy}^{(1)} = 0, \quad \sigma_y^{(1)} = \sigma^{\infty} \tag{6.61}$$

whereas the latter subproblem with the surface traction conditions

$$t_x^{(2)} = -\sigma^{\infty} n_x, \quad t_y^{(2)} = 0 \tag{6.62}$$

will be solved by means of the MFS. The superscript represents the first and second subproblems, respectively.

We first investigate the effect of the distance of the virtual and the actual boundaries (see Fig. 6.22). Here the ratio d/R is actually equal to $-\gamma$ defined in Eq. (1.48). Results at the point $(0, 0.4)$ are computed. It is noticed that at this point $\sigma_r = \sigma_y$, $\sigma_\theta = \sigma_x$, $\tau_{r\theta} = -\tau_{xy}$. Results in Fig. 6.23 indicate that the condition number of the coefficient matrix of the final linear solving system of equations increases when the radius of the virtual circular boundary decreases. But the accuracy of the stress component σ_x first increases and then reaches a relatively stable value. Next, the stress distributions along the y axis and the circular hole boundary are evaluated respectively to demonstrate the accuracy and convergence of the MFS. It is found in Fig. 6.24 that the results of stress components σ_r and σ_θ show good convergence as the number of boundary collocations increase, and 36 boundary collocations can produce excellent approximation to the analytical solutions. Moreover, the variation of hoop stress along the circular hole boundary is plotted in Fig. 6.25, from which we observe good agreement between the numerical and analytical results.

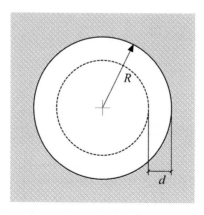

FIGURE 6.22 Illustration of position of the virtual boundary.

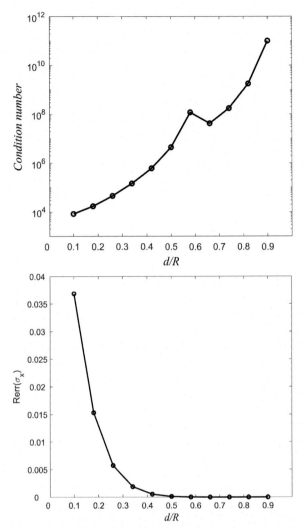

FIGURE 6.23 Effects of the distance of virtual and actual boundary on the condition number and the stress σ_x at point (0, 0.4).

Also, it is found that the maximum and minimum hoop stresses on the rim of a circular hole are about 3 MPa and −1 MPa, which are analytical stresses for this problem. The corresponding intensity factor is 3.

6.4 Solution procedure for inhomogeneous elasticity

For inhomogeneous elastic problems described in Eq. (6.18), which may be caused by the presence of body forces or temperature change in the domain, as

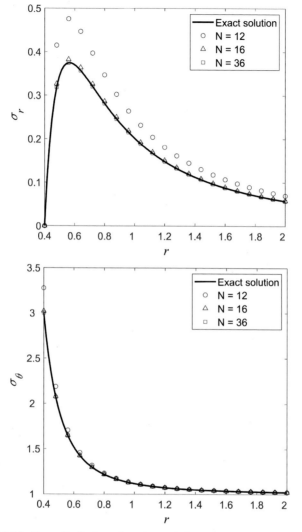

FIGURE 6.24 Stress distributions along the y axis for various numbers of collocations.

shown in Fig. 6.26, we first divide the solutions into two parts because of the linearity of the solution [7,8]:

$$u_i = u_i^p + u_i^h, \quad i = 1, 2 \tag{6.63}$$

where the particular part u_i^p satisfies

$$\left(\widetilde{\lambda} + \widetilde{\mu}\right) u_{k,ki}^p + \widetilde{\mu} u_{i,kk}^p + \widetilde{b}_i = 0 \tag{6.64}$$

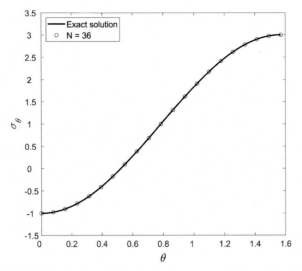

FIGURE 6.25   Hoop stress distribution along the circular hole boundary.

and the homogeneous part u_i^h satisfies

$$\left(\tilde{\lambda} + \tilde{\mu}\right)u_{k,ki}^h + \tilde{\mu}u_{i,kk}^h = 0 \tag{6.65}$$

Correspondingly, the boundary conditions Eq. (6.20) should be modified as

$$\begin{aligned}
u_i^h &= \bar{u}_i - u_i^p, \quad \text{on } \Gamma_u \\
t_i^h &= \bar{t}_i - t_i^p, \quad \text{on } \Gamma_t
\end{aligned} \tag{6.66}$$

Correspondingly, the strain and stress solutions can also be divided into the homogeneous and particular parts, i.e.,

$$\varepsilon_{ij} = \varepsilon_{ij}^p + \varepsilon_{ij}^h \tag{6.67}$$

$$\sigma_{ij} = \sigma_{ij}^p + \sigma_{ij}^h \tag{6.68}$$

where

$$\begin{aligned}
\varepsilon_{ij}^h &= \frac{1}{2}\left(u_{i,j}^h + u_{j,i}^h\right) \\
\varepsilon_{ij}^p &= \frac{1}{2}\left(u_{i,j}^p + u_{j,i}^p\right)
\end{aligned} \tag{6.69}$$

and

$$\begin{aligned}
\sigma_{ij}^h &= \tilde{\lambda}\delta_{ij}\varepsilon_{kk}^h + 2\tilde{\mu}\varepsilon_{ij}^h \\
\sigma_{ij}^p &= \tilde{\lambda}\delta_{ij}\varepsilon_{kk}^p + 2\tilde{\mu}\varepsilon_{ij}^p - \tilde{m}\delta_{ij}T
\end{aligned} \tag{6.70}$$

Further, the boundary traction is written as

$$t_i = t_i^p + t_i^h \tag{6.71}$$

where

$$
\begin{aligned}
t_i^h &= \sigma_{ij}^h n_j \\
t_i^p &= \sigma_{ij}^p n_j
\end{aligned} \tag{6.72}
$$

6.4.1 Particular solution

In this section, the RBFs will be used to derive the displacement particular solutions.

Firstly, the computing domain is represented by a finite number of nodes that may be arbitrarily dispersed in the computing domain, as indicated in Fig. 6.27. Then, the known values of body forces in Eq. (6.64) at arbitrary point $\mathbf{x}$ can be interpolated using the radial basis functions centered at these collocations:

$$\widetilde{b}_i(\mathbf{x}) = \sum_{m=1}^{M} \alpha_i^m \phi^m(\mathbf{x}) = \sum_{m=1}^{M} \alpha_l^m \delta_{li} \phi^m(\mathbf{x}), \quad \mathbf{x} \in \Omega \tag{6.73}$$

where M is the number of interpolating points possibly including interior and boundary collocation points, $\alpha_i^m (i = 1, 2; \ m = 1 \to M)$ are coefficients to be determined, and $\phi^m(\mathbf{x}) = \phi(\|\mathbf{x} - \mathbf{x}_m\|)$ is a set of RBFs centered at $\mathbf{x}_m \in \Omega$.

In matrix form, Eq. (6.73) is rewritten as

$$\widetilde{\mathbf{b}}(\mathbf{x}) = \mathbf{f}(\mathbf{x})\boldsymbol{\alpha} \tag{6.74}$$

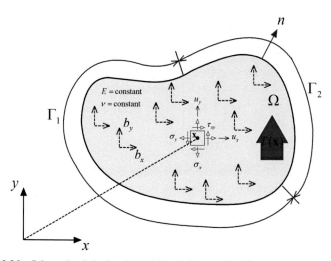

FIGURE 6.26 Schematic of elastic solids with body forces and subjected to temperature change.

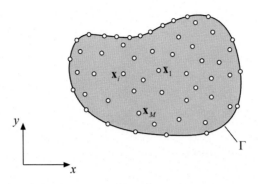

FIGURE 6.27 Collocation for particular solutions.

where

$$\widetilde{\mathbf{b}}(\mathbf{x}) = \begin{bmatrix} \widetilde{b}_1(\mathbf{x}) \\ \widetilde{b}_2(\mathbf{x}) \end{bmatrix}_{2\times 1} \tag{6.75}$$

$$\mathbf{f}(\mathbf{x}) = \begin{bmatrix} \mathbf{f}^1(\mathbf{x}) & \mathbf{f}^2(\mathbf{x}) & \cdots & \mathbf{f}^M(\mathbf{x}) \end{bmatrix}_{2\times 2M} \tag{6.76}$$

$$\boldsymbol{\alpha} = \begin{bmatrix} \boldsymbol{\alpha}^1 \\ \boldsymbol{\alpha}^2 \\ \vdots \\ \boldsymbol{\alpha}^M \end{bmatrix}_{2M\times 1} \tag{6.77}$$

with the following submatrices:

$$\mathbf{f}^m(\mathbf{x}) = \begin{bmatrix} \phi^m(\mathbf{x}) & 0 \\ 0 & \phi^m(\mathbf{x}) \end{bmatrix}_{2\times 2}, \boldsymbol{\alpha}^m = \begin{bmatrix} \alpha_1^m \\ \alpha_2^m \end{bmatrix}_{2\times 1}, \quad m = 1, 2, ..., M \tag{6.78}$$

Letting $\mathbf{x}$ take the M collocations in turn, we obtain the following linear system of equations in matrix form:

$$\begin{bmatrix} \mathbf{f}(\mathbf{x}_1) \\ \mathbf{f}(\mathbf{x}_2) \\ \vdots \\ \mathbf{f}(\mathbf{x}_M) \end{bmatrix} \boldsymbol{\alpha} = \begin{bmatrix} \widetilde{\mathbf{b}}(\mathbf{x}_1) \\ \widetilde{\mathbf{b}}(\mathbf{x}_2) \\ \vdots \\ \widetilde{\mathbf{b}}(\mathbf{x}_M) \end{bmatrix} \tag{6.79}$$

from which the unknown coefficients α_i^m can be evaluated.

Next, the particular solution u_i^p can be written in a similar form as

$$u_i^p(\mathbf{x}) = \sum_{m=1}^M \alpha_l^m \Phi_{li}^m(\mathbf{x}), \quad \mathbf{x} \in \Omega \tag{6.80}$$

where Φ_{li}^m is a corresponding set of approximate particular solutions.

Eq. (6.80) can be rewritten in matrix form:

$$\mathbf{u}^p = \mathbf{\Phi}\boldsymbol{\alpha} \tag{6.81}$$

where

$$\mathbf{u}^p(\mathbf{x}) = \begin{bmatrix} u_1^p(\mathbf{x}) \\ u_2^p(\mathbf{x}) \end{bmatrix} \tag{6.82}$$

$$\mathbf{\Phi} = \begin{bmatrix} \mathbf{\Phi}^1(\mathbf{x}) & \mathbf{\Phi}^2(\mathbf{x}) & \cdots & \mathbf{\Phi}^M(\mathbf{x}) \end{bmatrix}_{2 \times 2M} \tag{6.83}$$

with

$$\mathbf{\Phi}^m(\mathbf{x}) = \begin{bmatrix} \Phi_{11}^m(\mathbf{x}) & \Phi_{21}^m(\mathbf{x}) \\ \Phi_{12}^m(\mathbf{x}) & \Phi_{22}^m(\mathbf{x}) \end{bmatrix}_{2 \times 2}, \quad m = 1, 2, ..., M \tag{6.84}$$

Because the particular solution u_i^p is required to satisfy Eq. (6.64), substitution of Eqs. (6.80) and (6.73) into Eq. (6.64) yields the following equations in tensor notation:

$$\left(\tilde{\lambda} + \tilde{\mu}\right)\Phi_{ki,kl}^m(\mathbf{x}) + \tilde{\mu}\Phi_{li,kk}^m(\mathbf{x}) = -\delta_{li}\phi^m(\mathbf{x}), \quad \mathbf{x} \in \Omega \tag{6.85}$$

Consequently, the derivative of displacement in terms of spatial variable can give the following strain particular solution:

$$\varepsilon_{ij}^p(\mathbf{x}) = \frac{1}{2}\left(u_{i,j}^p + u_{j,i}^p\right) = \sum_{m=1}^M \alpha_l^m T_{lij}^m(\mathbf{x}), \quad \mathbf{x} \in \Omega \tag{6.86}$$

from which the stress particular solution can be written as

$$\sigma_{ij}^p(\mathbf{x}) = \tilde{\lambda}\delta_{ij}\varepsilon_{kk}^p + 2\tilde{\mu}\varepsilon_{ij}^p - \tilde{m}\delta_{ij}T = \sum_{m=1}^M \alpha_l^m S_{lij}^m(\mathbf{x}) - \tilde{m}\delta_{ij}T, \quad \mathbf{x} \in \Omega \tag{6.87}$$

where

$$T_{lij}^m(\mathbf{x}) = \frac{1}{2}\left[\Phi_{il,j}^m(\mathbf{x}) + \Phi_{jl,i}^m(\mathbf{x})\right], \quad m = 1, 2, ..., M \tag{6.88}$$

$$S_{lij}^m(\mathbf{x}) = \left[\tilde{\lambda}\delta_{ij}\Phi_{lk,k}^m(\mathbf{x}) + \tilde{\mu}T_{lij}^m(\mathbf{x})\right], \quad m = 1, 2, ..., M \tag{6.89}$$

Similarly, Eqs. (6.86) and (6.87) can be rewritten in matrix form:

$$\boldsymbol{\varepsilon}^p(\mathbf{x}) = \widehat{\mathbf{B}}(\mathbf{x})\boldsymbol{\alpha}, \quad \mathbf{x} \in \Omega \tag{6.90}$$

$$\boldsymbol{\sigma}^p(\mathbf{x}) = \widehat{\mathbf{S}}(\mathbf{x})\boldsymbol{\alpha} - \mathbf{T}, \quad \mathbf{x} \in \Omega \tag{6.91}$$

where

$$\widehat{\mathbf{B}}(\mathbf{x}) = \begin{bmatrix} \widehat{\mathbf{B}}^1(\mathbf{x}) & \widehat{\mathbf{B}}^2(\mathbf{x}) & \cdots & \widehat{\mathbf{B}}^M(\mathbf{x}) \end{bmatrix}_{3 \times 2M} \tag{6.92}$$

$$\widehat{\mathbf{S}}(\mathbf{x}) = \left[\widehat{\mathbf{s}}^1(\mathbf{x}) \quad \widehat{\mathbf{s}}^2(\mathbf{x}) \quad \cdots \quad \widehat{\mathbf{s}}^M(\mathbf{x}) \right]_{3 \times 2M} \tag{6.93}$$

$$\mathbf{T} = \begin{bmatrix} \widetilde{m}T \\ \widetilde{m}T \\ 0 \end{bmatrix} \tag{6.94}$$

and

$$\widehat{\mathbf{B}}^m(\mathbf{x}) = \begin{bmatrix} B_{111}^m(\mathbf{x}) & B_{211}^m(\mathbf{x}) \\ \widehat{B}_{122}^m(\mathbf{x}) & \widehat{B}_{222}^m(\mathbf{x}) \\ B_{112}^m(\mathbf{x}) & B_{212}^m(\mathbf{x}) \end{bmatrix}_{3 \times 2}, \quad m = 1, 2, ..., M \tag{6.95}$$

$$\widehat{\mathbf{S}}^m(\mathbf{x}) = \begin{bmatrix} S_{111}^m(\mathbf{x}) & S_{211}^m(\mathbf{x}) \\ S_{122}^m(\mathbf{x}) & S_{222}^m(\mathbf{x}) \\ S_{112}^m(\mathbf{x}) & S_{212}^m(\mathbf{x}) \end{bmatrix}_{3 \times 2}, \quad m = 1, 2, ..., M \tag{6.96}$$

Then, the boundary particular traction can be given by

$$\begin{aligned} t_i^p(\mathbf{x}) = \sigma_{ij}^p n_j &= \sum_{m=1}^{M} \alpha_l^m S_{lij}^m(\mathbf{x}) n_j - \widetilde{m} \delta_{ij} T n_j \\ &= \sum_{m=1}^{M} \alpha_l^m S_{lij}^m(\mathbf{x}) n_j - \widetilde{m} T n_i, \quad \mathbf{x} \in \Gamma \end{aligned} \tag{6.97}$$

or

$$\mathbf{t}^p(\mathbf{x}) = \mathbf{A}\boldsymbol{\sigma}^p(\mathbf{x}) = \mathbf{A}\widehat{\mathbf{S}}(\mathbf{x})\boldsymbol{\alpha} - \mathbf{A}\mathbf{T} = \widehat{\mathbf{T}}(\mathbf{x})\boldsymbol{\alpha} - \mathbf{A}\mathbf{T}, \quad \mathbf{x} \in \Gamma \tag{6.98}$$

The matrix $\mathbf{A}$ consists of the components of unit outward normal vector to the boundary Γ and is given in Eq. (2.37).

Now, the key point is how to determine the approximate particular solutions Φ_{li}^m from Eq. (6.85) if the radial basis function ϕ^m is given. To do this, the approximate particular solutions Φ_{li}^m in the context of linear plane strain elasticity can expressed by the Galerkin–Papkovich vector F_{li} as shown [9]:

$$\Phi_{li} = \frac{1-\nu}{\mu} F_{li,mm} - \frac{1}{2\mu} F_{mi,ml} \tag{6.99}$$

Substituting Eq. (6.99) into Eq. (6.85), we obtain

$$\begin{aligned} \frac{\mu}{1-2\nu}\Phi_{kl,ki} + \mu\Phi_{li,kk} &= \frac{1-\nu}{1-2\nu} F_{ki,mmkl} - \frac{1}{2(1-2\nu)} F_{mi,mkkl} \\ &\quad + (1-\nu)F_{li,mmkk} - \frac{1}{2} F_{mi,mlkk} \\ &= \frac{1-\nu}{1-2\nu} F_{ki,mmkl} - \frac{1-\nu}{(1-2\nu)} F_{mi,mkkl} + (1-\nu)F_{li,mmkk} \\ &= (1-\nu)F_{li,mmkk} = (1-\nu)\nabla^4 F_{li} \end{aligned} \tag{6.100}$$

Then, we have

$$\nabla^4 F_{li} = F_{li,mmkk} = -\frac{1}{1-\nu}\delta_{li}\phi \tag{6.101}$$

It is worth noting that the superscript "m" in the approximate particular solutions Φ_{li}^m and the RBF ϕ^m is removed in the derivation of Eqs. (6.99)−(6.101), for the sake of convenience.

In particular, if a continuous function Θ, like the RBF in the study, applies only with respect to the radial variable r, that is, $\Theta = \Theta(r)$, the biharmonic equation of such a function in two-dimensional polar coordinates is rewritten as

$$\nabla^4 \Theta(r) = \frac{1}{r}\frac{d}{dr}\left(r\frac{d}{dr}\left(\frac{1}{r}\frac{d}{dr}\left(r\frac{d\Theta}{dr}\right)\right)\right) \tag{6.102}$$

With this radial feature of biharmonic operator, Eq. (6.101) can be further written as

$$\frac{1}{r}\frac{d}{dr}\left(r\frac{d}{dr}\left(\frac{1}{r}\frac{d}{dr}\left(r\frac{dF_{li}(r)}{dr}\right)\right)\right) = -\frac{1}{1-\nu}\delta_{li}\phi(r) \tag{6.103}$$

Integrating Eq. (6.103) we can obtain the Galerkin−Papkovich vector $F_{li}(r)$ when a radial basis function $\phi(r)$ is given, and then the approximate particular solution $\Phi_{li}(r)$ can be derived by Eq. (6.99).

In this chapter, two popular RBFs are used for performing the RBF interpolation: (1) the piecewise smooth power spline (PS), $\phi^m = r^{2n-1}$, $n = 1$, 2, 3, ..., also known as conical spline; (2) the thin plate spline (TPS), $\phi^m = r^{2n}\ln r$, $n = 1, 2, 3, ...$, also called the Duchon spline.

For the PS basis under the plane strain state, integrating Eq. (6.101) and then substituting the Galerkin−Papkovich vector F_{li} into Eq. (6.99) yields

$$\Phi_{li}^m = -\frac{1}{2\mu(1-\nu)}\frac{1}{(2n+1)^2(2n+3)}r^{2n+1}\left(A_1\delta_{li} + A_2 r_{,l}r_{,i}\right) \tag{6.104}$$

where

$$\begin{aligned}A_1 &= (4n+5) - 2\nu(2n+3)\\ A_2 &= -(2n+1)\end{aligned} \tag{6.105}$$

Then, from Eq. (6.89) we have

$$S_{lij}^m = -\frac{1}{2(1-\nu)}\frac{r^{2n}}{(2n+1)^2(2n+3)}\left[C_1\delta_{ij}r_{,l} + C_2 r_{,i}r_{,j}r_{,l} + C_3\left(\delta_{il}r_{,j} + \delta_{lj}r_{,i}\right)\right] \tag{6.106}$$

where

$$\begin{aligned}C_1 &= 2(2n+1)[(2n+3)\nu - 1]\\ C_2 &= -2(4n^2 - 1)\\ C_3 &= (2n+1)[4(n+1) - 2\nu(2n+3)]\end{aligned} \tag{6.107}$$

Similarly, for the TPS basis under the plane strain state, we have

$$\Phi_{li}^{m} = -\frac{1}{32\mu(1-v)} \frac{r^{2n+2}}{(n+1)^{3}(n+2)^{2}} \left(A_{1}\delta_{il} + A_{2}r_{,i}r_{,l}\right) \qquad (6.108)$$

where

$$
\begin{aligned}
A_{1} &= -(8n^{2}+29n+27)+8v(n+2)^{2} \\
&\quad +2(n+1)(n+2)[4n+7-4v(n+2)]\ln r \qquad (6.109) \\
A_{2} &= 2(n+1)(2n+3)-4(n+1)^{2}(n+2)\ln r
\end{aligned}
$$

Then, from Eq. (6.89) we have

$$S_{lij}^{m} = -\frac{1}{8(1-v)} \frac{r^{2n+1}}{(n+1)^{2}(n+2)^{2}} \left[C_{1}\delta_{ij}r_{,l} + C_{2}r_{,i}r_{,j}r_{,l} + C_{3}\left(\delta_{il}r_{,j} + \delta_{lj}r_{,i}\right)\right]$$

$$(6.110)$$

where

$$
\begin{aligned}
C_{1} &= 2n+3-2v(n+2)^{2}+2(n+1)(n+2)(2vn+4v-1)\ln r \\
C_{2} &= 2(n^{2}-2)-4n(n+1)(n+2)\ln r \\
C_{3} &= -(2n^{2}+6n+5)+2v(n+2)^{2} \\
&\quad -2(n+1)(n+2)[-(2n+3)+2v(n+2)]\ln r
\end{aligned} \qquad (6.111)
$$

In the preceding derivation, the first and second derivatives of the approximate particular solutions in tensor form are needed. All these derivatives can be found in Appendix C for completeness.

As a result, the displacement and stress particular solutions at arbitrary point $\mathbf{x}$ in the computing domain can be evaluated through Eqs. (6.80) and (6.87). Once this step is completed, the corresponding homogeneous solutions can be obtained by the standard method of fundamental solutions in the following sections.

6.4.2 Homogeneous solution

To obtain the strongly approximated solutions of the homogeneous partial differential Eq. (6.65) under the modified boundary conditions Eq. (6.66), N fictitious source points $\mathbf{y}_{n}$ ($n = 1, 2, \ldots, N$) are placed on a virtual boundary outside the domain. Moreover, assume that at each source point there is a pair of fictitious point loads φ_{1}^{i}, φ_{2}^{i} along the x and y directions, respectively.

According to the basic concept of MFS, the homogenous displacement and stress fields at arbitrary field points $\mathbf{x} \in \Omega$ inside the domain or on its boundary can be approximated by a linear combination of fundamental solutions in terms of those fictitious sources outside the domain of interest:

$$u_{i}^{h}(\mathbf{x}) = \sum_{n=1}^{N} \varphi_{l}^{n} u_{li}^{*}(\mathbf{x}, \mathbf{y}_{n}) \qquad (6.112)$$

$$\sigma_{ij}^h(\mathbf{x}) = \sum_{n=1}^{N} \varphi_l^n \sigma_{lij}^*(\mathbf{x}, \mathbf{y}_n) \tag{6.113}$$

Then, the homogenous boundary traction is given by

$$t_i^h(\mathbf{x}) = \sum_{n=1}^{N} \varphi_l^n t_{li}^*(\mathbf{x}, \mathbf{y}_n) \tag{6.114}$$

In matrix form, Eqs. (6.112)–(6.114) can be rewritten as

$$\mathbf{u}^h(\mathbf{x}) = \mathbf{U}(\mathbf{x})\boldsymbol{\varphi} \tag{6.115}$$

$$\boldsymbol{\sigma}^h(\mathbf{x}) = \mathbf{S}(\mathbf{x})\boldsymbol{\varphi} \tag{6.116}$$

$$\mathbf{t}^h(\mathbf{x}) = \mathbf{T}(\mathbf{x})\boldsymbol{\varphi} \tag{6.117}$$

where

$$\mathbf{u}^h(\mathbf{x}) = \left\{ \begin{array}{c} u_1^h(\mathbf{x}) \\ u_2^h(\mathbf{x}) \end{array} \right\}, \quad \boldsymbol{\sigma}^h(\mathbf{x}) = \left\{ \begin{array}{c} \sigma_{11}^h(\mathbf{x}) \\ \sigma_{22}^h(\mathbf{x}) \\ \sigma_{12}^h(\mathbf{x}) \end{array} \right\}, \quad \mathbf{t}^h(\mathbf{x}) = \left\{ \begin{array}{c} t_1^h(\mathbf{x}) \\ t_2^h(\mathbf{x}) \end{array} \right\} \tag{6.118}$$

Because the displacement and stress solutions in Eqs. (6.112) and (6.113) exactly satisfy the homogeneous governing Eq. (6.65), only the specific boundary conditions Eq. (6.66) are needed to be satisfied in the MFS.

To satisfy the boundary conditions point by point, a series of collocations $\mathbf{x}_i$ ($i = 1 \rightarrow L$) are chosen from the real boundary. Then we have

$$\sum_{n=1}^{N} \varphi_l^n u_{li}^*(\mathbf{x}_i^u, \mathbf{y}_n) = \bar{u}_i(\mathbf{x}_i^u) - u_i^p(\mathbf{x}_i^u), \quad i = 1 \rightarrow L_1, \ \mathbf{x}_i^u \in \Gamma_1$$

$$\sum_{n=1}^{N} \varphi_l^n t_{li}^*(\mathbf{x}_i^t, \mathbf{y}_n) = \bar{t}_i(\mathbf{x}_i^t) - t_i^p(\mathbf{x}_i^t), \quad i = 1 \rightarrow L_2, \ \mathbf{x}_i^t \in \Gamma_t \tag{6.119}$$

or in matrix form

$$\begin{bmatrix} \mathbf{U}(\mathbf{x}_1^u) \\ \vdots \\ \mathbf{U}(\mathbf{x}_{L_1}^u) \\ \hline \mathbf{T}(\mathbf{x}_1^t) \\ \vdots \\ \mathbf{T}(\mathbf{x}_{L_2}^t) \end{bmatrix}_{2L \times 2N} \boldsymbol{\varphi} = \begin{bmatrix} \bar{\mathbf{u}}(\mathbf{x}_1^u) - \mathbf{u}^p(\mathbf{x}_1^u) \\ \vdots \\ \bar{\mathbf{u}}(\mathbf{x}_{L_1}^u) - \mathbf{u}^p(\mathbf{x}_{L_1}^u) \\ \hline \bar{\mathbf{t}}(\mathbf{x}_1^t) - \mathbf{t}^p(\mathbf{x}_1^t) \\ \vdots \\ \bar{\mathbf{t}}(\mathbf{x}_{L_2}^t) - \mathbf{t}^p(\mathbf{x}_{L_2}^t) \end{bmatrix}_{2L \times 1} \tag{6.120}$$

which finally produces the following linear system of discrete equations in terms of the unknown coefficients in matrix form:

$$\mathbf{H}_{2L \times 2N} \, \boldsymbol{\varphi}_{2N \times 1} = \mathbf{P}_{2L \times 1} \tag{6.121}$$

Solving Eq. (6.121) can determine the unknown coefficients numerically. After this, the homogeneous displacements and stresses at arbitrary point in the domain can be evaluated by Eqs. (6.112) and (6.113).

6.4.3 Approximated full solution

Once the homogeneous and particular displacement, strain, and stress solutions at an arbitrary point in the domain are determined, the full solutions can naturally be written as

$$u_i(\mathbf{x}) = u_i^p(\mathbf{x}) + u_i^h(\mathbf{x}) \tag{6.122}$$

$$\varepsilon_{ij}(\mathbf{x}) = \varepsilon_{ij}^p(\mathbf{x}) + \varepsilon_{ij}^h(\mathbf{x}) \tag{6.123}$$

$$\sigma_{ij}(\mathbf{x}) = \sigma_{ij}^p(\mathbf{x}) + \sigma_{ij}^h(\mathbf{x}) \tag{6.124}$$

6.4.4 Results and discussion

In this section, two inhomogeneous examples with available analytical solutions are considered to validate the meshless approach present in this book.

6.4.4.1 Rotating disk with high speed

Consider a compact disk with inner radius $a = 7.5$ mm, outer radius $b = 60$ mm, and thickness of $h = 1.2$ mm, rotating about the z-axis with constant angular velocity $\omega = 10^4$ r/min, as shown in Fig. 6.28. The material properties used are $E = 1.6 \times 10^4$ MPa, $\nu = 0.3$, and $\rho = 2.2 \times 10^{-9}$ mg/mm^3.

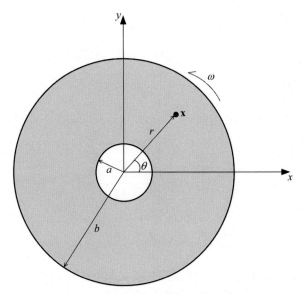

FIGURE 6.28 Schematic of rotating disk with high speed.

In the absence of the weight of the disk, just the inertial force caused by rotation contributes to the generalized body forces. For such a case, the generalized body forces $\widetilde{b}_i$ $(i = 1, 2)$ are linear functions in terms of spatial coordinate x_i, that is,

$$\widetilde{b}_i = \rho \omega^2 x_i \qquad (6.125)$$

If there are no forces acting on all boundaries, and the tangential displacement on the inner boundary is set to zero, the analytical solutions (plane stress) in polar coordinates are available for this problem [2]:

$$u_r = \frac{1}{E}\left[(1 - \nu)Cr - (1 + \nu)C_1\frac{1}{r} - \frac{1 - \nu^2}{8}\rho\omega^2 r^3\right] \qquad (6.126)$$

$$u_\theta = 0$$

and

$$\sigma_r = C + C_1\frac{1}{r^2} - \frac{3 + \nu}{8}\rho\omega^2 r^2$$

$$\sigma_\theta = C - C_1\frac{1}{r^2} - \frac{1 + 3\nu}{8}\rho\omega^2 r^2 \qquad (6.127)$$

$$\tau_{r\theta} = 0$$

where

$$C = \frac{3 + \nu}{8}\rho\omega^2\left(a^2 + b^2\right), \quad C_1 = -\frac{3 + \nu}{8}\rho\omega^2 a^2 b^2 \qquad (6.128)$$

Fig. 6.29 plots the variation of analytical radial and hoop stresses along the radial direction, from which it is found that the hoop stress caused by rotation

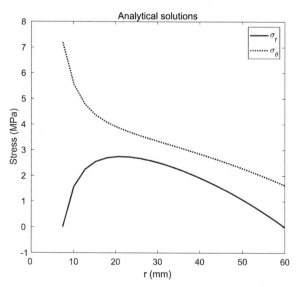

FIGURE 6.29 Analytical solutions of normal stress components along the radial direction.

is a tensile stress and has its maximum value at the inner boundary. Although the caused radial stress is also tensile, its value is obviously smaller than that of the hoop stress, and its maximum stress appears in the region close to the inner boundary. Therefore, the high-speed rotating disc may experience damage at the inner boundary.

To investigate the convergence and accuracy of the present meshless method, numerical results of radial and hoop stresses with different boundary collocations are provided in Figs. 6.30 and 6.31. It is observed from Figs. 6.30 and 6.31

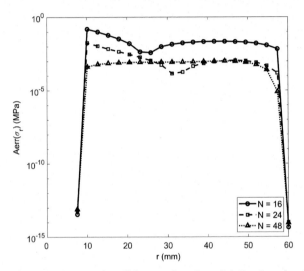

FIGURE 6.30 Absolute error in radial stress along the radial direction when $M = 75$.

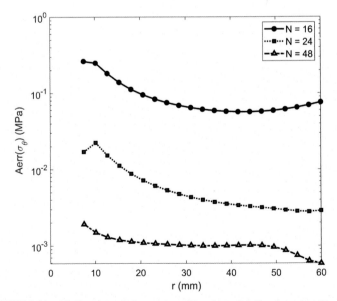

FIGURE 6.31 Absolute error in hoop stress along the radial direction when $M = 75$.

TABLE 6.1 Illustration of convergence of the meshless method when $N = 48$.

M	45	75	135	195
Arerr(σ_r)	1.2×10^{-3}	1.7×10^{-4}	4.2×10^{-5}	3.0×10^{-5}
Arerr(σ_θ)	1.6×10^{-3}	2.4×10^{-4}	6.3×10^{-5}	4.6×10^{-5}

that the accuracy of stress components increases as the number of boundary collocations increase. Moreover, it is found that good accuracy can be achieved by a few boundary collocations. Additionally, the numerical results in Table 6.1 clearly show the convergence of the present meshless method for increased M and constant $N = 48$.

6.4.4.2 Symmetric thermoelastic problem in a long cylinder

Consider a long cylinder with inner radius a and outer radius b, as shown in Fig. 6.32. It is assumed that a radial temperature change applies in the domain as

$$T(r) = \frac{\ln \dfrac{r}{b}}{\ln \dfrac{a}{b}} T_0 \tag{6.129}$$

where T_0 is a constant.

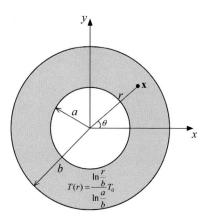

FIGURE 6.32 Schematic of thermoelasticity in a long thick-walled cylinder.

In the absence of body forces, the analytical stress components caused by temperature change are given as shown next [2]:

$$\sigma_r = -\frac{E\alpha T_0}{2(1-\nu)}\left(\frac{\ln\frac{b}{r}}{\ln\frac{b}{a}} - \frac{\frac{b^2}{r^2}-1}{\frac{b^2}{a^2}-1}\right)$$

$$\sigma_\theta = -\frac{E\alpha T_0}{2(1-\nu)}\left(\frac{\ln\frac{b}{r}-1}{\ln\frac{b}{a}} + \frac{\frac{b^2}{r^2}+1}{\frac{b^2}{a^2}-1}\right)$$

(6.130)

$$\tau_{r\theta} = 0$$

In the computation, $a = 3$ m, $b = 6$ m, $E = 69$ GPa, $\nu = 0.33$, $\alpha = 23 \times 10^{-6}/°C$, and $T_0 = 10°C$. Fig. 6.33 plots the analytical variations in the radial and hoop stresses along the radial direction, from which it is found that the hoop stress caused by the temperature change varies more dramatically in the cylinder than the radial stress. The maximum hoop stress happening at the outer boundary is close to 10 MPa (tensile stress), and the minimum value, occurring at the inner boundary, is about -15 MPa (compressive stress).

To investigate the convergence and accuracy of the present meshless method, the effect of the number of RBF interpolation points M is first investigated. Table 6.2 tabulates the relative error of stresses for various M,

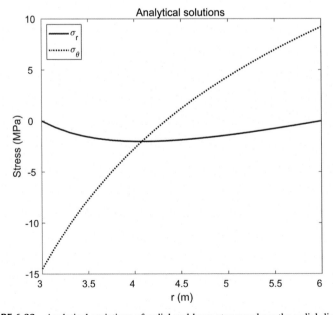

FIGURE 6.33 Analytical variations of radial and hoop stresses along the radial direction.

TABLE 6.2 Illustration of convergence of the meshless method when $N = 48$.

M	75	105	195	260
$Arerr(\sigma_r)$	0.0464	0.0225	0.0138	0.0050
$Arerr(\sigma_\theta)$	0.0502	0.0311	0.0229	0.0096

TABLE 6.3 Illustration of convergence of the meshless method when $M = 105$.

N	24	48
$Arerr(\sigma_r)$	0.0638	0.0225
$Arerr(\sigma_\theta)$	0.0531	0.0311

from which it is found that the accuracy of numerical results improves as M increases. Moreover, when $M = 105$, the results in Table 6.3 indicate that the relative error of stresses also decreases as more boundary collocations are employed. Finally, the numerical results of radial and hoop stresses with $N = 48$ and $M = 260$ are plotted along the radial direction in Figs. 6.34 and 6.35, from which it is clearly observed that the present meshless method produces almost same results as the analytical solutions.

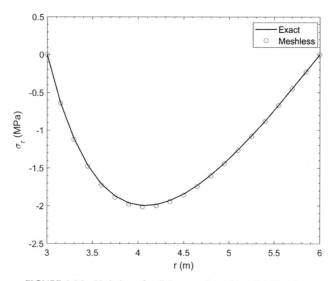

FIGURE 6.34 Variation of radial stress along the radial direction.

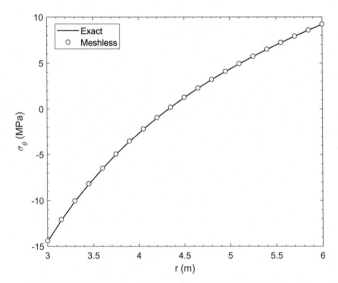

FIGURE 6.35 Variation of hoop stress along the radial direction.

6.5 Further analysis for functionally graded solids

In the preceding analysis, only homogeneous solids with constant elastic properties are considered. In fact, the elastic parameter is permitted to be changed in the functionally graded materials (FGMs) [10]. In this section, the concept of FGMs and the basic formulations of thermoelasticity in functionally graded elastic solids are reviewed first, and then the meshless solution procedure is established for thermoelastic analysis in FGMs [11]. Finally, some numerical examples are tested to validate the present meshless method.

6.5.1 Concept of functionally graded material

FGMs are omnipresent in nature. From naturally growing bamboo with radial density gradient [12,13] to bone with naturally optimized spongy trabecular structure [14,15], graded microstructures make these materials have excellent strength and stiffness to weight ratio to meet their expected service requirements under the complex loading conditions to which they are subjected. Fig. 6.36 illustrates the microstructure in the radial direction of bamboo cross-section, and it is clearly seen that the bamboo fiber cells have varying radial density, so it can resist bending and torsion more efficiently.

In contrast to pure materials having constant physical properties, as well as composite materials which are made up of one or more materials combined in solid states to achieve distinct physical properties, FGMs have spatially smoothly varying composition or microstructure, which means that their physical properties change continuously in terms of spatial dimensions. Owing

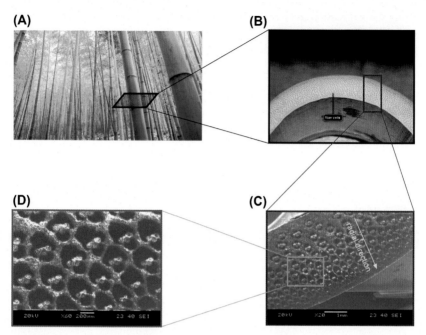

FIGURE 6.36 Graded microstructure in radial direction of bamboo material. (A) Bamboo stalks; (B) cross-section; (C) radial distribution of fiber cells in SEM picture; (D) enlarged picture.

to such smooth transition from one material constituent to another constituent, an FGM can eliminate the existence of mismatched interfaces in composite materials at which delamination failure always occurs. Therefore, FGMs can be applied in aerospace, automobile, bioengineering, [10] etc.

Currently, there are different ways to fabricate FGMs [10]. For a thin functionally graded coating such as thermal barrier coating, the physical or chemical vapor deposition (PVD/CVD) can be used to produce it, while bulk FGM can be produced using the powder metallurgy technique. For both thin and bulk FGMs, it is generally assumed that the material property, such as density, elastic modulus, thermal conductivity, etc., changes along one direction following a rule. For example, Fig. 6.37 displays the constituent transition

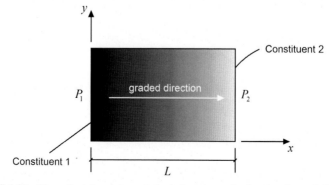

FIGURE 6.37 Theoretical description of physical property of a functionally graded material.

along the x direction in an FGM from the constituent 1 with property P_1 to the constituent 2 with property P_2. To describe such smooth variation of physical property P of the FGM, a continuous function can be defined as

$$P = P(x) \tag{6.131}$$

Typically, two continuous variations are assumed in practice. One is the power-law assumption, and another is the exponential assumption.

For the power-law assumption, the physical property of an FGM based on the Voigt model is written as follows [16]:

$$P(x) = [1 - V(x)]P_1 + V(x)P_2 \tag{6.132}$$

where $V(x)$ denotes the volume fraction of the constituent 2 and changes in power form

$$V(x) = \left(\frac{x}{L}\right)^n \tag{6.133}$$

in which L is the length of graded direction, and n is the given graded parameter.

Then, the substitution of Eq. (6.133) into Eq. (6.132) yields

$$P(x) = \left[1 - \left(\frac{x}{L}\right)^n\right]P_1 + \left(\frac{x}{L}\right)^n P_2 \tag{6.134}$$

To demonstrate the power-law variation of the physical property, Fig. 6.38 gives several curves of the physical property of the FGM for various graded parameter n. It is resulting from Fig. 6.38 that $n \neq 1$ gives nonlinear variation of the physical property $P(x)$, while $n = 1$ corresponds to the linear variation.

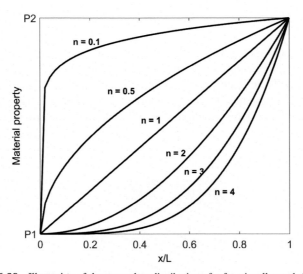

FIGURE 6.38 Illustration of the power-law distribution of a functionally graded material.

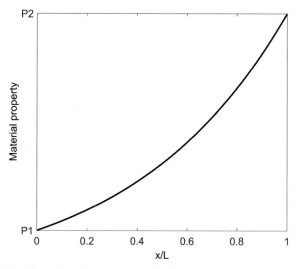

FIGURE 6.39   Illustration of the power-law distribution of a functionally graded material.

Besides, it is observed that too small or too big a value of the graded parameter n produces a steeper curve.

On the other hand, if the exponential variation is assumed, the physical property of an FGM can be given by

$$P(x) = e^{\frac{x}{L}\eta}P_1 \tag{6.135}$$

from which the graded parameter η can be evaluated by the known values of the physical property at the left and right ends, i.e.,

$$\eta = \ln\frac{P_2}{P_1} \tag{6.136}$$

Then, substituting Eq. (6.136) into Eq. (6.135) gives

$$P(x) = e^{\frac{x}{L}\ln\frac{P_2}{P_1}}P_1 \tag{6.137}$$

whose variation curve is plotted in Fig. 6.39.

6.5.2 Thermomechanical systems in FGMs

Different from the homogeneous elastic materials, which have constant elastic properties, the elastic properties, e.g., Young's modulus, of the FGMs change in terms of spatial position along a certain direction, so the FGMs can be regarded as a special case of general heterogeneous materials. Due to this feature, the governing equations of FGMs are more complex than those corresponding to homogeneous material. In this section, the formulations and solution procedure provided subsequently can be used to depict general thermomechanics problems in two-dimensional isotropic heterogeneous elastic solids.

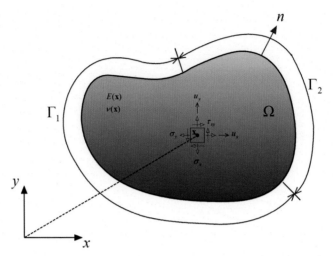

FIGURE 6.40 Schematic of two-dimensional isotropic elastic solid with variable elastic parameters.

Without loss of generality, let us consider a general two-dimensional domain Ω bounded by the boundary Γ. The material involved is isotropic and heterogeneously elastic. If the Cartesian coordinates $\mathbf{x} = \{x_1, x_2\}^T$ or $\{x, y\}^T$ are used to describe infinitesimal static deformations depicting by the field variables u_1 and u_2 in the domain Ω, the elastic properties, i.e., Young's modulus, of the heterogeneous material are functions of the spatial variable $\mathbf{x}$, as indicated in Fig. 6.40.

6.5.2.1 Strain-displacement relationship

In the linear elastic theory, the strain-displacement relationship is independent of material definition, thus, the same as that in isotropic and homogeneous elasticity, if the displacements are small enough that the square and product of its derivatives are negligible; then the relation of Cauchy strains ε_{ij} and displacements u_i can be written as

$$\varepsilon_{ij}(\mathbf{x}) = \frac{1}{2}\left(u_{i,j}(\mathbf{x}) + u_{j,i}(\mathbf{x})\right) \tag{6.138}$$

where $i, j = 1, 2$ denote, respectively, the x_1 and x_2 directions. Here and thereafter, the Cartesian tensor notation is adopted for the sake of convenience. The comma in the equations indicates a space derivative, e.g., $u_{i,j} = \partial u_i / \partial x_j$.

In matrix form, the strain-displacement relationship can be given by

$$\varepsilon = \mathbf{L}\mathbf{u} \tag{6.139}$$

where

$$
\mathbf{L} = \begin{bmatrix} \dfrac{\partial}{\partial x} & 0 \\[2mm] 0 & \dfrac{\partial}{\partial y} \\[2mm] \dfrac{\partial}{\partial y} & \dfrac{\partial}{\partial x} \end{bmatrix}
\tag{6.140}
$$

6.5.2.2 Constitutive equations

For an isotropic heterogeneous elastic material with variable elastic parameters, the constitutive equations related to stresses and strains are stated in the following tensor form:

$$
\sigma_{ij}(\mathbf{x}) = \widetilde{\lambda}(\mathbf{x})\delta_{ij}\varepsilon_{kk}(\mathbf{x}) + 2\widetilde{\mu}(\mathbf{x})\varepsilon_{ij}(\mathbf{x}) - \widetilde{m}(\mathbf{x})\delta_{ij}T(\mathbf{x})
\tag{6.141}
$$

where σ_{ij} denotes the components of Cauchy stress tensor, and T denotes the temperature change the material experiences, that is, the final temperature minus the original temperature. The last term in the right-hand side represents the temperature effect of the material; for instance, if the change in temperature is positive, we have thermal expansion, and if negative, thermal contraction.

Besides, due to the heterogeneous definition of the material, the material parameters in Eq. (6.141) are functions of spatial coordinates:

$$
\widetilde{\lambda}(\mathbf{x}) = \frac{2\widetilde{\nu}(\mathbf{x})}{1 - 2\widetilde{\nu}(\mathbf{x})}\widetilde{\mu}(\mathbf{x}), \quad \widetilde{\mu}(\mathbf{x}) = \frac{\widetilde{E}(\mathbf{x})}{2(1 + \widetilde{\nu}(\mathbf{x}))}, \quad \widetilde{m}(\mathbf{x}) = \frac{\widetilde{\alpha}(\mathbf{x})\widetilde{E}(\mathbf{x})}{1 - 2\widetilde{\nu}(\mathbf{x})}
\tag{6.142}
$$

where $\widetilde{E}(\mathbf{x})$, $\widetilde{\nu}(\mathbf{x})$, $\widetilde{\alpha}(\mathbf{x})$ have different values for solids in the plane stress and plane strain states, i.e.,

$$
\widetilde{E}(\mathbf{x}) = E(\mathbf{x}), \quad \widetilde{\nu}(\mathbf{x}) = \nu(\mathbf{x}), \quad \widetilde{\alpha}(\mathbf{x}) = \alpha(\mathbf{x})
\tag{6.143}
$$

for the plane strain state and

$$
\widetilde{E}(\mathbf{x}) = \frac{1 + 2\nu(\mathbf{x})}{(1 + \nu(\mathbf{x}))^2}E(\mathbf{x}), \quad \widetilde{\nu}(\mathbf{x}) = \frac{\nu(\mathbf{x})}{1 + \nu(\mathbf{x})}, \quad \widetilde{\alpha}(\mathbf{x}) = \frac{1 + \nu(\mathbf{x})}{1 + 2\nu(\mathbf{x})}\alpha(\mathbf{x})
\tag{6.144}
$$

for the plane stress state. Here, the elastic modulus $E(\mathbf{x})$, the Poisson's ratio $\nu(\mathbf{x})$, and the linear coefficient of thermal expansion $\alpha(\mathbf{x})$ are functions of space coordinates $\mathbf{x}$, respectively. Besides, in Eq. (6.141), the same subscript appearing twice in an equation represents summation, thus $\varepsilon_{kk} = \varepsilon_{11} + \varepsilon_{22}$ stands by the strain invariable.

In matrix form, we have

$$\boldsymbol{\sigma}(\mathbf{x}) = \mathbf{D}(\mathbf{x})\boldsymbol{\varepsilon}(\mathbf{x}) - \widetilde{m}(\mathbf{x})T(\mathbf{x})\mathbf{I} \tag{6.145}$$

where

$$\mathbf{D}(\mathbf{x}) = \begin{bmatrix} \widetilde{\lambda}(\mathbf{x}) + 2\widetilde{\mu}(\mathbf{x}) & \widetilde{\mu}(\mathbf{x}) & 0 \\ \widetilde{\mu}(\mathbf{x}) & \widetilde{\lambda}(\mathbf{x}) + 2\widetilde{\mu}(\mathbf{x}) & 0 \\ 0 & 0 & \widetilde{\mu}(\mathbf{x}) \end{bmatrix} \tag{6.146}$$

and

$$\mathbf{I} = \begin{bmatrix} 1 & 0 & 0 \\ 0 & 1 & 0 \\ 0 & 0 & 0 \end{bmatrix} \tag{6.147}$$

6.5.2.3 Static equilibrium equations

Same as that for 2D homogeneous elastic solids, the equilibrium equations for 2D heterogeneously elastic solids can be written in the following form using tensor notations:

$$\sigma_{ij,j} + b_i = 0, \quad \text{in } \Omega \tag{6.148}$$

where b_i is the components of body force per unit volume. In Eq. (6.148), the first term in the left-hand side involves the partial derivative to the spatial variable and the summation notation for repeated subscript, e.g.,

$$\sigma_{ij,j} = \frac{\partial \sigma_{ij}}{\partial x_j} = \frac{\partial \sigma_{i1}}{\partial x_1} + \frac{\partial \sigma_{i2}}{\partial x_2}$$

In matrix form, we have

$$\mathbf{L}^{\mathrm{T}}\boldsymbol{\sigma} + \mathbf{b} = 0 \tag{6.149}$$

Furthermore, substituting Eq. (6.138) into Eq. (6.141) and then into Eq. (6.148) yields the second-order partial differential equation in terms of displacement components:

$$\left(\widetilde{\lambda} + \widetilde{\mu}\right)u_{k,ki} + \widetilde{\mu}u_{i,kk} + \widetilde{\lambda}_{,i}u_{k,k} + \widetilde{\mu}_{,k}\left(u_{i,k} + u_{k,i}\right) - \widetilde{m}T_{,i} - \widetilde{m}_{,i}T + b_i = 0 \tag{6.150}$$

or

$$\mathbf{L}^{\mathrm{T}}\left(\mathbf{D}(\mathbf{x})\mathbf{L}\mathbf{u}(\mathbf{x}) - \widetilde{m}(\mathbf{x})T(\mathbf{x})\mathbf{I}\right) + \mathbf{b} = 0 \tag{6.151}$$

Compared to the equilibrium Eq. (6.18) for 2D isotropic and homogeneously elastic solids, Eq. (6.150) includes additional terms related to the spatial derivative of the material parameters, since these material parameters are functions of space coordinates for FGMs.

6.5.2.4 Boundary conditions

The complete system defining the boundary value problem for two-dimensional elastic solids should consist of the strain-displacement Eq. (6.138), the constitutive Eq. (6.141), the equilibrium Eq. (6.148), and the specific boundary conditions, which generally include the following displacement and traction conditions applied to the boundary of the domain:

$$u_i = \bar{u}_i, \qquad \text{on } \Gamma_u$$
$$t_i = \sigma_{ij}n_j = \bar{t}_i, \quad \text{on } \Gamma_t \tag{6.152}$$

where $\bar{u}_i$ is the prescribed displacements on Γ_u and $\bar{t}_i$ the given tractions on Γ_t. Γ_u and Γ_t are complementary parts of the boundary Γ. n_j represents the direction cosines of the unit outward normal to the boundary.

6.5.3 Solution procedure for FGMs

6.5.3.1 Analog equation method

For the thermoelastic Eq. (6.150) describing displacement responses in general heterogeneous media, the fundamental solutions are difficult to obtain in a closed form. However, we can circumvent this obstacle by the indirect way, which is called the analog equation method (AEM) by J.T. Katsikadelis [17]. In this method, the basic concept is to enforce the known standard partial differential operator to the sought field solution to establish an equivalent governing equation to the original one mathematically. Here, the unknown field is displacement, as depicted in Eq. (6.150). From the viewpoint of mathematics, the displacement fields must be in terms of space coordinates, regardless of the particular forms of elastic properties and loading types. So we can introduce an equivalent elastic system written as

$$\left(\widehat{\lambda} + \widehat{\mu}\right)u_{k,ki} + \widehat{\mu}u_{i,kk} + \widehat{b}_i = 0 \tag{6.153}$$

to replace the original governing Eq. (6.150). In Eq. (6.153), $\widehat{\lambda}$, $\widehat{\mu}$ are elastic constants of a fictitious isotropic homogeneous solid, and $\widehat{b}_i$ is the fictitious body forces induced by unknown the displacement solution. In practical computation, the elastic constants $\widehat{\lambda}$ and $\widehat{\mu}$ can be chosen as the average values of that of the two material constituents in the FGM of interest; i.e., for the plane strain state, they can be written as

$$\widehat{\lambda} = \frac{2v_a}{1 - 2v_a}G_a, \quad \widehat{\mu} = G_a = \frac{E_a}{2(1 + v_a)}, \quad m = \frac{\alpha_a E_a}{1 - 2v_a} \tag{6.154}$$

where

$$E_a = \frac{E_1 + E_2}{2}, \quad v_a = \frac{v_1 + v_2}{2}, \quad \alpha_a = \frac{\alpha_1 + \alpha_2}{2} \tag{6.155}$$

E_i, v_i, and α_i ($i = 1, 2$) are elastic modulus, Poisson's ratio, and thermal expansion coefficient of the ith material constituent, respectively.

In the following procedure, we will derive the general displacement solutions of Eq. (6.153) using the RBF approximation and the MFS.

Obviously, the displacement solution in the new linear partial differential system Eq. (6.153) can be divided into two parts:

$$u_i(\mathbf{x}) = u_i^p(\mathbf{x}) + u_i^h(\mathbf{x}) \tag{6.156}$$

where u_i^p is the particular solution part satisfying

$$\left(\widehat{\lambda} + \widehat{\mu}\right) u_{k,ki}^p + \widehat{\mu} u_{i,kk}^p + \widehat{b}_i = 0 \tag{6.157}$$

regardless of the boundary conditions, and the complementary part u_i^h is the homogeneous solution that satisfies

$$\left(\widehat{\lambda} + \widehat{\mu}\right) u_{k,ki}^h + \widehat{\mu} u_{i,kk}^h = 0 \tag{6.158}$$

Next, we will use the RBF approximation and the MFS to derive the expressions of two parts.

6.5.3.2 Particular solution

By collocating nodes in the computing domain, as indicated in Fig. 6.27, the fictitious body forces in Eq. (6.157) at arbitrary point $\mathbf{x}$ can be interpolated as

$$\widehat{b}_i(\mathbf{x}) = \sum_{m=1}^{M} \phi^m(\mathbf{x}) \alpha_i^m = \sum_{m=1}^{M} \delta_{li} \phi^m(\mathbf{x}) \alpha_l^m \tag{6.159}$$

where M is the number of interpolating points in the computing domain, α_i^m are coefficients to be determined, and $\phi^m(\mathbf{x}) = \phi(\|\mathbf{x} - \mathbf{x}_m\|)$ is a set of RBFs centered at $\mathbf{x}_m \in \Omega$.

Similarly, the particular solution u_i^p is also approximated by means of the same coefficient set

$$u_i^p(\mathbf{x}) = \sum_{m=1}^{M} \Phi_{li}^m(\mathbf{x}) \alpha_l^m \tag{6.160}$$

where $\Phi_{li}^m(\mathbf{x})$ is a corresponding set of approximate particular solutions, which is related to the given radial basis functions $\phi^m(\mathbf{x})$ by

$$\left(\widehat{\lambda} + \widehat{\mu}\right) \Phi_{lk,ki}^m(\mathbf{x}) + \widehat{\mu} \Phi_{li,kk}^m(\mathbf{x}) = -\delta_{li} \phi^m(\mathbf{x}) \tag{6.161}$$

For convenience, Eq. (6.160) can be rewritten in matrix form:

$$\mathbf{u}^p(\mathbf{x}) = \mathbf{\Phi}(\mathbf{x})\boldsymbol{\alpha} \tag{6.162}$$

where

$$\mathbf{u}^p(\mathbf{x}) = \begin{bmatrix} u_1^p(\mathbf{x}) \\ u_2^p(\mathbf{x}) \end{bmatrix} \tag{6.163}$$

$$\boldsymbol{\Phi}(\mathbf{x}) = \left[\boldsymbol{\Phi}^1(\mathbf{x}) \quad \boldsymbol{\Phi}^2(\mathbf{x}) \quad \cdots \quad \boldsymbol{\Phi}^M(\mathbf{x}) \right]_{2 \times 2M} \tag{6.164}$$

with

$$\boldsymbol{\Phi}^m(\mathbf{x}) = \begin{bmatrix} \Phi_{11}^m(\mathbf{x}) & \Phi_{21}^m(\mathbf{x}) \\ \Phi_{12}^m(\mathbf{x}) & \Phi_{22}^m(\mathbf{x}) \end{bmatrix}_{2 \times 2}, \quad m = 1, 2, \ldots, M \tag{6.165}$$

6.5.3.3 Homogeneous solution

To obtain approximated solutions of the homogeneous Eq. (6.158), N fictitious source points $\mathbf{y}_n$ ($n = 1, 2, \ldots, N$) located on the virtual boundary outside the domain are selected (Fig. 6.5). Moreover, assume that at each source point there is a pair of fictitious point loads φ_1^i, φ_2^i along x and y direction, respectively. According to the main construct of MFS, the approximated homogeneous displacement fields at arbitrary field point $\mathbf{x}$ in the domain can be expressed as a linear combination of fundamental solutions of the new equivalent Eq. (6.158) with linear isotropic and homogenous material definition, that is,

$$u_i^h(\mathbf{x}) = \sum_{n=1}^{N} u_{li}^*(\mathbf{x}, \mathbf{y}_n) \varphi_l^n, \quad \forall \mathbf{x} \in \Omega, \ \mathbf{y}_n \notin \Omega \tag{6.166}$$

in which the displacement fundamental solution $u_{li}^*(\mathbf{x}, \mathbf{y})$ satisfies the following equivalent Cauchy-Navier partial differential equations subject to the unit concentrated force along the l-direction at the source point $\mathbf{y}$ and with the constant elastic properties $\widehat{\lambda}$ and $\widehat{\mu}$:

$$\left(\widehat{\lambda} + \widehat{\mu} \right) u_{lk,ki}^*(\mathbf{x}, \mathbf{y}) + \widehat{\mu} u_{li,kk}^*(\mathbf{x}, \mathbf{y}) = -\delta_{\mathbf{xy}} e_{li} \tag{6.167}$$

whose solutions under the plane strain state are expressed as

$$u_{li}^*(\mathbf{x}, \mathbf{x}_s) = \frac{1}{8\pi G_\alpha \left(1 - \widehat{v} \right)} \left[(3 - 4v_\alpha) \delta_{li} \ln \frac{1}{r} + r_{,l} r_{,i} \right] \tag{6.168}$$

In matrix form, Eq. (6.166) can be rewritten as

$$\mathbf{u}^h(\mathbf{x}) = \mathbf{U}(\mathbf{x}) \boldsymbol{\varphi} \tag{6.169}$$

where

$$\mathbf{u}^h(\mathbf{x}) = \left\{ \begin{matrix} u_1^h(\mathbf{x}) \\ u_2^h(\mathbf{x}) \end{matrix} \right\} \tag{6.170}$$

$$\mathbf{U}(\mathbf{x}) = \left[\mathbf{u}^*(\mathbf{x}, \mathbf{y}_1) \quad \mathbf{u}^*(\mathbf{x}, \mathbf{y}_2) \quad \cdots \quad \mathbf{u}^*(\mathbf{x}, \mathbf{y}_N) \right]_{2 \times 2N} \tag{6.171}$$

and

$$\mathbf{u}^*(\mathbf{x}, \mathbf{y}_n) = \begin{bmatrix} u^*_{11}(\mathbf{x}, \mathbf{y}_n) & u^*_{21}(\mathbf{x}, \mathbf{y}_n) \\ u^*_{12}(\mathbf{x}, \mathbf{y}_n) & u^*_{22}(\mathbf{x}, \mathbf{y}_n) \end{bmatrix}_{2 \times 2}, \quad n = 1, 2, \dots, N \quad (6.172)$$

It is apparent that the approximate solution in Eq. (6.166) completely satisfies the new equivalent homogeneous Eq. (6.158) in the domain based on the definition of the fundamental solutions, if the fact that source point $\mathbf{y}_n$ and field point $\mathbf{x}$ do not overlap holds in the computation.

6.5.3.4 Approximated full solution

From Eq. (6.156), the full solutions of displacement components are written as the sum of the particular and homogeneous solutions:

$$u_i(\mathbf{x}) = \sum_{n=1}^{N} u^*_{li}(\mathbf{x}, \mathbf{y}_n)\varphi^n_l + \sum_{m=1}^{M} \Phi^m_{li}(\mathbf{x})\alpha^m_l, \quad \mathbf{x} \in \Omega, \ \mathbf{y}_n \notin \Omega \quad (6.173)$$

Differentiating Eq. (6.173) yields

$$u_{i,j}(\mathbf{x}) = \sum_{n=1}^{N} u^*_{li,j}(\mathbf{x}, \mathbf{y}_n)\varphi^n_l + \sum_{m=1}^{M} \alpha^m_l \Phi^m_{li,j}(\mathbf{x}), \quad \mathbf{x} \in \Omega, \ \mathbf{y}_n \notin \Omega \quad (6.174)$$

$$u_{i,jk}(\mathbf{x}) = \sum_{n=1}^{N} u^*_{li,jk}(\mathbf{x}, \mathbf{y}_n)\varphi^n_l + \sum_{m=1}^{M} \alpha^m_l \Phi^m_{li,jk}(\mathbf{x}), \quad \mathbf{x} \in \Omega, \ \mathbf{y}_n \notin \Omega \quad (6.175)$$

Consequently, the strain and stress components can be obtained by using the strain-displacement relationship Eq. (6.138) and the real constitutive Eq. (6.141):

$$\begin{aligned} \varepsilon_{ij}(\mathbf{x}) &= \frac{1}{2}\left(u_{i,j} + u_{j,i}\right) \\ &= \sum_{n=1}^{N} \varepsilon^*_{lij}(\mathbf{x}, \mathbf{y}_n)\varphi^n_l + \sum_{m=1}^{M} B^m_{lij}(\mathbf{x})\alpha^m_l, \quad \mathbf{x} \in \Omega, \ \mathbf{y}_n \notin \Omega \end{aligned} \quad (6.176)$$

$$\begin{aligned} \sigma_{ij}(\mathbf{x}) &= \tilde{\lambda}(\mathbf{x})\delta_{ij}\varepsilon_{kk}(\mathbf{x}) + 2\tilde{\mu}(\mathbf{x})\varepsilon_{ij}(\mathbf{x}) - \tilde{m}(\mathbf{x})\delta_{ij}T(\mathbf{x}) \\ &= \sum_{n=1}^{N} \sigma^*_{lij}(\mathbf{x}, \mathbf{y}_n)\varphi^n_l + \sum_{m=1}^{M} S^m_{lij}(\mathbf{x})\alpha^m_l - \tilde{m}\delta_{ij}T(\mathbf{x}), \quad \mathbf{x} \in \Omega, \ \mathbf{y}_n \notin \Omega \end{aligned} \quad (6.177)$$

where

$$\begin{aligned} \varepsilon^*_{lij}(\mathbf{x}) &= \frac{1}{2}\left(u^*_{li,j}(\mathbf{x}) + u^*_{lj,i}(\mathbf{x})\right) \\ B^m_{lij}(\mathbf{x}) &= \frac{1}{2}\left(\Phi^m_{li,j}(\mathbf{x}) + \Phi^m_{lj,i}(\mathbf{x})\right) \end{aligned} \quad (6.178)$$

$$\sigma^*_{lij}(\mathbf{x}) = \tilde{\lambda}(\mathbf{x})\delta_{ij}u^*_{lk,k}(\mathbf{x}) + \tilde{\mu}(\mathbf{x})\left(u^*_{li,j}(\mathbf{x}) + u^*_{lj,i}(\mathbf{x})\right)$$
$$S^m_{lij}(\mathbf{x}) = \tilde{\lambda}(\mathbf{x})\delta_{ij}\Phi^m_{lk,k}(\mathbf{x}) + \tilde{\mu}(\mathbf{x})\left(\Phi^m_{li,j}(\mathbf{x}) + \Phi^m_{lj,i}(\mathbf{x})\right)$$

(6.179)

Furthermore, the traction components on the boundary Γ can be given by

$$t_i(\mathbf{x}) = \sigma_{ij}(\mathbf{x})n_j$$
$$= \sum_{n=1}^{N} t^*_{li}(\mathbf{x},\mathbf{y}_n)\varphi^n_l + \sum_{m=1}^{M} P^m_{li}(\mathbf{x})\alpha^m_l - \tilde{m}n_i T(\mathbf{x}), \quad \mathbf{x}\in\Gamma,\ \mathbf{y}_n\notin\Omega$$

(6.180)

where

$$t^*_{li}(\mathbf{x},\mathbf{y}_n) = \sigma^*_{lij}(\mathbf{x},\mathbf{y}_n)n_j$$
$$P^m_{li}(\mathbf{x}) = S^m_{lij}(\mathbf{x})n_j$$

(6.181)

In matrix form, the previous expressions of displacement, strain, stress, and traction can be rewritten as

$$\mathbf{u}(\mathbf{x}) = \mathbf{U}(\mathbf{x})\boldsymbol{\varphi} + \boldsymbol{\Phi}(\mathbf{x})\boldsymbol{\alpha} = \begin{bmatrix} \mathbf{U}(\mathbf{x}) & \boldsymbol{\Phi}(\mathbf{x}) \end{bmatrix}\begin{bmatrix}\boldsymbol{\varphi}\\\boldsymbol{\alpha}\end{bmatrix}, \quad \mathbf{x}\in\Omega \qquad (6.182)$$

$$\boldsymbol{\varepsilon}(\mathbf{x}) = \mathbf{B}(\mathbf{x})\boldsymbol{\varphi} + \widehat{\mathbf{B}}(\mathbf{x})\boldsymbol{\alpha} = \begin{bmatrix} \mathbf{B}(\mathbf{x}) & \widehat{\mathbf{B}}(\mathbf{x}) \end{bmatrix}\begin{bmatrix}\boldsymbol{\varphi}\\\boldsymbol{\alpha}\end{bmatrix}, \quad \mathbf{x}\in\Omega \qquad (6.183)$$

$$\boldsymbol{\sigma}(\mathbf{x}) = \mathbf{S}(\mathbf{x})\boldsymbol{\varphi} + \widehat{\mathbf{S}}(\mathbf{x})\boldsymbol{\alpha} - \mathbf{T}(\mathbf{x}) = \begin{bmatrix} \mathbf{S}(\mathbf{x}) & \widehat{\mathbf{S}}(\mathbf{x}) \end{bmatrix}\begin{bmatrix}\boldsymbol{\varphi}\\\boldsymbol{\alpha}\end{bmatrix} - \mathbf{T}(\mathbf{x}), \quad \mathbf{x}\in\Omega$$

(6.184)

$$\mathbf{t}(\mathbf{x}) = \mathbf{T}(\mathbf{x})\boldsymbol{\varphi} + \widehat{\mathbf{T}}(\mathbf{x})\boldsymbol{\alpha} = \begin{bmatrix} \mathbf{T}(\mathbf{x}) & \widehat{\mathbf{T}}(\mathbf{x}) \end{bmatrix}\begin{bmatrix}\boldsymbol{\varphi}\\\boldsymbol{\alpha}\end{bmatrix}, \quad \mathbf{x}\in\Gamma \qquad (6.185)$$

where the matrices $\mathbf{U}(\mathbf{x})$, $\boldsymbol{\Phi}(\mathbf{x})$, $\mathbf{B}(\mathbf{x})$, $\widehat{\mathbf{B}}(\mathbf{x})$, $\mathbf{S}(\mathbf{x})$, $\widehat{\mathbf{S}}(\mathbf{x})$, $\mathbf{T}(\mathbf{x})$, $\mathbf{T}(\mathbf{x})$, and $\widehat{\mathbf{T}}(\mathbf{x})$ are the same as those given in Sections 6.3 and 6.4; however, the kernels in them should be replaced with Eqs. (6.178), (6.179), and (6.181).

6.5.3.5 Construction of solving equations

Once the explicit expressions of the approximations Eqs. (6.182)–(6.185) are obtained, we can determine all unknowns in them by enforcing them to satisfy the original partial differential Eq. (6.150) and the boundary conditions Eq. (6.152). To do this, making Eq. (6.182) satisfy the original governing Eq. (6.150) at M interpolation points dispersed in the domain and simultaneously making Eqs. (6.182) and (6.185) satisfy the boundary conditions Eq. (6.152) at N boundary nodes will produce

$$\begin{bmatrix} \mathbf{L}^{\mathrm{T}}\left(\mathbf{D}(\mathbf{x})\mathbf{L}\mathbf{U}(\mathbf{x})\right) & \mathbf{L}^{\mathrm{T}}\left(\mathbf{D}(\mathbf{x})\mathbf{L}\boldsymbol{\Phi}(\mathbf{x})\right) \\ \hline \mathbf{U}(\mathbf{x}) & \boldsymbol{\Phi}(\mathbf{x}) \\ \mathbf{T}(\mathbf{x}) & \widehat{\mathbf{T}}(\mathbf{x}) \end{bmatrix}\begin{bmatrix}\boldsymbol{\varphi}\\\boldsymbol{\alpha}\end{bmatrix} = \begin{bmatrix} \mathbf{L}^{\mathrm{T}}\left(\tilde{m}(\mathbf{x})T(\mathbf{x})\mathbf{I}\right) - \mathbf{b} \\ \hline \bar{\mathbf{u}} \\ \bar{\mathbf{t}} \end{bmatrix} \qquad (6.186)$$

which finally leads to a $2(M + N) \times 2(M + N)$ linear algebraic system of discrete equations in matrix form for the determination of unknown coefficients:

$$\mathbf{HA} = \mathbf{B} \tag{6.187}$$

where the unknown vector

$$\mathbf{A} = \begin{bmatrix} \boldsymbol{\varphi} \\ \boldsymbol{\alpha} \end{bmatrix} \tag{6.188}$$

For simplicity, the temperature distribution used in the following computation is taken in an analytical form, rather than numerical solution obtained from a boundary value problem of heat conduction. It should be, however, mentioned that the idea of MFS with RBF also can be used to obtain the numerical distribution of temperature in an FGM, and the detailed procedure is documented in Chapter 8.

6.5.4 Numerical experiments

In this section, three examples of FGMs subjected to mechanical or thermal loads are considered to assess the proposed algorithm. In all three examples, except for Poisson's ratio, the material properties vary exponentially or according to a power law. This is a reasonable assumption, since variation on the Poisson's ratio is usually small compared with that of other properties [11,16,18,19]. Finally, to assess the accuracy and convergence of the approximation, the average relative error $Arerr(\zeta)$ of any field variable ζ defined in Eq. (1.51) is adopted.

6.5.4.1 Functionally graded hollow circular plate under radial internal pressure

Consider a hollow circular functionally graded plate as shown in Fig. 6.41 with inner radius $a = 5$ mm and outer radius $b = 10$ mm under internal radial

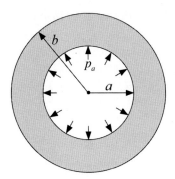

FIGURE 6.41 Configuration of hollow circular functionally graded plate under internal pressure.

pressure. Suppose the plate is graded along the radial direction in power form so that elastic modulus

$$E(r) = E_0 \left(\frac{r}{a}\right)^{\eta}$$

(6.189)

where η is the power number. For $\eta > 0$, the Young's modulus increases as the radius r increases. As $\eta = 0$, the problem is reduced to the analysis of homogeneous media. Analytical solutions of stress components for the case of plane stress state are given in closed form Ref. [20]:

$$\sigma_r = -\frac{a^{-\frac{\eta}{2}} r^{-1-\frac{k}{2}+\frac{\eta}{2}}}{b^k - a^k} a^{1+\frac{k}{2}} (b^k - r^k) p_a$$

$$\sigma_\theta = \frac{a^{-\frac{\eta}{2}} r^{-1-\frac{k}{2}+\frac{\eta}{2}}}{b^k - a^k} \left[\frac{(2 + k\nu - \eta\nu)r^k}{k - \eta + 2\nu} - \frac{(-2 + k\nu + \eta\nu)b^k}{k + \eta - 2\nu} \right] a^{1+\frac{k}{2}} p_a$$

(6.190)

with $k = \sqrt{\eta^2 + 4 - 4\eta\nu}$.

In the practical computation, Poisson's ratio, elastic modulus at the internal surface, as well as internal pressure, respectively, are assumed to be $\nu = 0.3$, $E(a) = 200$ GPa, and $p_a = 50$ MPa. Figs. 6.42 and 6.43 display the convergent performance of the proposed meshless method when the PS basis function r^3 is used. It is found from Figs. 6.42 and 6.43 that the accuracy increases with an increase in M or N.

To investigate the variation of radial and hoop stresses along the radial direction for various graded parameters η, 32 boundary nodes and 140 interior interpolation points are used. Comparisons between analytical solutions and numerical results are shown in Figs. 6.44 and 6.45. It is found that regardless of the value of η, radial stress increases monotonously from the inner to the

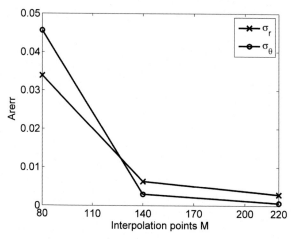

FIGURE 6.42 Convergent performance versus M ($\eta = 2$, N $= 32$).

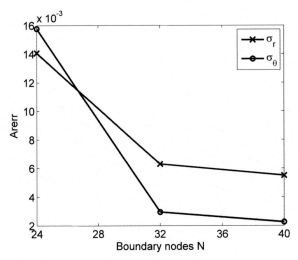

FIGURE 6.43 Convergent performance versus N ($\eta = 2$, M = 140).

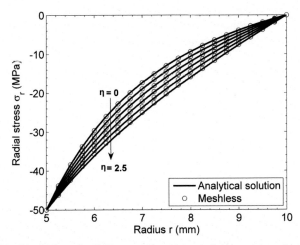

FIGURE 6.44 Distribution of radial stress with various graded parameters when 32 boundary nodes and 140 interior interpolation points are used.

outer surface, whereas hoop stress does not. As η increases, the value of radial stress decreases at any point in the cylinder, except for the points on the boundary, whereas the maximum hoop stress occurs on the inner surface when $\eta = 0$ and on the outer surface when $\eta = 4$. The variation in the hoop stress looks like rotation around a center when η increases. It is also found that the variation in hoop stress in FGMs becomes worse when η increases. Therefore, to avoid material instability, the graded parameter should be smaller than specific values.

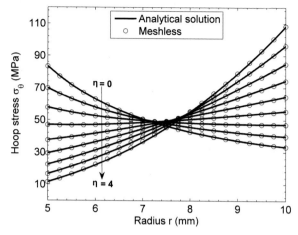

FIGURE 6.45 Distribution of hoop stress with various graded parameters when 32 boundary nodes and 140 interior interpolation points are used.

In this example, the effects of the types and orders of RBF are also tested for the case of the high graded parameter $\eta = 2$. Figs. 6.46 and 6.47 show the average relative error distributions caused by the PS and TPS basis functions. It is evident that a higher order of RBF does not always result in better accuracy. The calculation indicates that r^3 and r^5 in PS, and $r^2 lnr$ and $r^4 lnr$ in TPS seem to be able to produce relatively high accuracy in this example. Moreover, TPS has better accuracy than PS. Therefore, $r^4 lnr$ is used in the remaining computation.

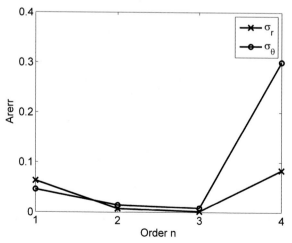

FIGURE 6.46 Effect of orders of PS when 32 boundary nodes and 220 interior interpolation points are used for the case of $\eta = 2$.

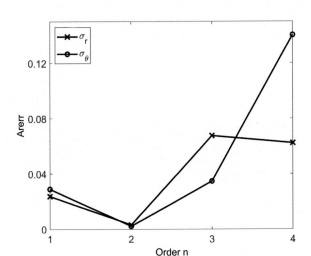

FIGURE 6.47 Effect of orders of TPS when 32 boundary nodes and 220 interior interpolation points are used for the case of $\eta = 2$.

6.5.4.2 Functionally graded elastic beam under sinusoidal transverse load

An elastic beam as shown in Fig. 6.48 is considered in the second example, which is made of two-phase Al/SiC composite. The elastic modulus varying exponentially in the z direction is given by $E(z) = E_0 e^{nz}$. The left and right end faces of the functionally graded beam are assumed to be simply supported, so

$$w(0, z) = w(L, z) = 0$$
$$t_x(0, z) = t_x(L, z) = 0 \tag{6.191}$$

The top surface of the beam is assumed to be free of mechanical force and the bottom surface is subjected to a distributed load P, as shown in Fig. 6.48:

$$t_x(x, 0) = t_x(x, h) = 0$$
$$t_z(x, 0) = p, \quad t_z(x, h) = 0 \tag{6.192}$$

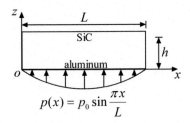

FIGURE 6.48 A functionally graded beam subjected to symmetric sinusoidal transverse loading.

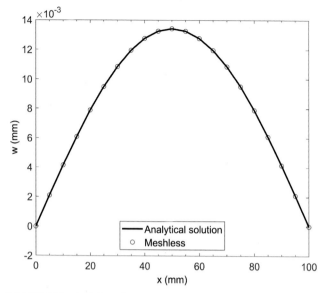

FIGURE 6.49 Variation of transverse displacement along the line $z = h/2$.

The problem is solved under a plane strain assumption with the length $L = 100$ mm and thickness $h = 40$ mm. The material properties of aluminum and SiC are, respectively, $E_{Al} = 70$ GPa and $E_{SiC} = 427$ GPa, thus the graded parameter is

$$\eta = \frac{\ln(E_{SiC}/E_{Al})}{h} \tag{6.193}$$

In the computation, the maximum transverse load P_0 is assumed to be equal to 10 MPa. A total of 34 boundary nodes and 169 interior interpolation points are selected in the analysis. Figs. 6.49 and 6.50 respectively show the variation of transverse displacement and stress components along the horizontal line $z = h/2$, and Fig. 6.51 plots the stress components along the vertical line $x = L/5$. Good agreement can be observed between the numerical results and exact solutions [18]. Furthermore, the shapes of cross-sections after deformation are provided in Fig. 6.52, from which it can be seen that for smaller ratios of thickness and length, for example, $h/L = 1/10$, the cross-section approximately maintains plane after deformation. This phenomenon demonstrates the validity of the cross-section assumption in classic thin beam bending theory.

6.5.4.3 Symmetrical thermoelastic problem in a long functionally graded cylinder

In the last example, a thick hollow cylinder with the same geometries and mechanical boundary conditions as given in Fig. 6.41 is considered. The same

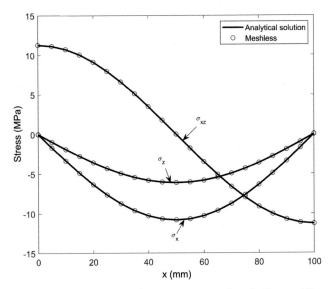

FIGURE 6.50 Variations of stress components along the line $z = h/2$.

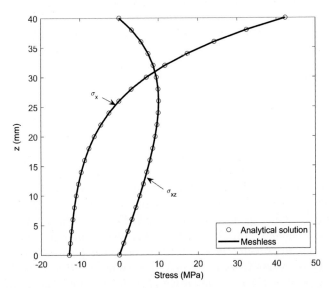

FIGURE 6.51 Variation of stress components along the cross-section $x = L/5$.

power-law assumptions are used to define the elastic modulus and coefficient of thermal expansion, that is,

$$E(r) = E_0 \left(\frac{r}{a}\right)^\eta, \quad \alpha(r) = \alpha_0 \left(\frac{r}{a}\right)^\eta \tag{6.194}$$

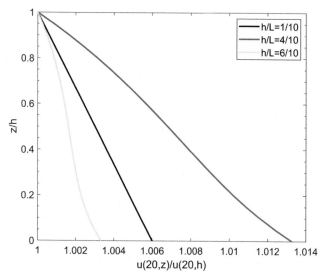

FIGURE 6.52   Shape of transverse cross-section after deformation with various ratio of thickness and length.

The temperature change in the entire domain is given in a closed form:

$$T = \begin{cases} \dfrac{T_a(b^{-\eta} - r^{-\eta}) + T_b(r^{-\eta} - a^{-\eta})}{b^{-\eta} - a^{-\eta}} & \text{for } \eta \neq 0 \\[3mm] \dfrac{T_a \ln\dfrac{b}{r} + T_b \ln\dfrac{r}{a}}{\ln\dfrac{b}{a}} & \text{for } \eta = 0 \end{cases} \tag{6.195}$$

with $T_a = T(a)$ and $T_b = T(b)$.

The two-phase aluminum/ceramic FGM is examined here. The metal aluminum constituent is arranged on the inner surface, while the ceramic constituent is on the outer surface. The related material properties are $E_{Al} = 70$ GPa, $\alpha_{Al} = 1.2 \times 10^{-6}/°C$, $E_{ceramic} = 151$ GPa, and $\alpha_{ceramic} = 2.59 \times 10^{-6}/°C$. Poisson's ratio is taken to be $v = 0.3$. The inner and outer boundary temperature changes, respectively, are $T_a = 10°C$ and $T_b = 0°C$.

Analytical solutions of displacements and stresses for the case of plane strain state are provided by Jabbari et al. [21]. The results in Figs. 6.53 and 6.54 show good agreement between the analytical solutions and the numerical results in FGM and homogeneous material, which corresponds to $\eta = 0$. Furthermore, we again find that after graded treatment, the maximum value of hoop stress decreases from 82.6—53 MPa. Additionally, the radial displacement in FGM also decreases, compared to the response in homogeneous media. Since the value of radial displacement is very small, radial deformation can be neglected in practical analysis.

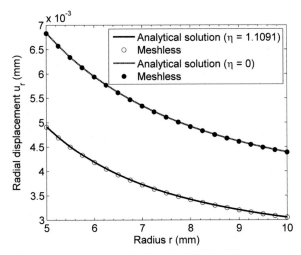

FIGURE 6.53 Radial displacement distributions in FGM and homogeneous material with $N = 32$ and $M = 220$.

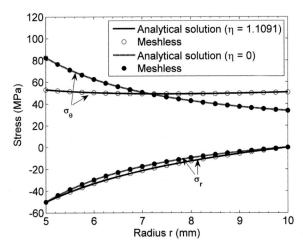

FIGURE 6.54 Stress distributions in FGM and homogeneous material with $N = 32$ and $M = 220$.

6.6 Remarks

By coupling the RBF interpolation and the MFS, a mixed meshless method is presented in this chapter for solving two-dimensional linear elastic problems with unknown body forces that may be caused by the gravity, rotation, or temperature change in the domain. Results indicate that the present method has good accuracy and convergence for such problems.

Subsequently, the present meshless method is extended to perform thermoelastic analysis in the functionally graded solids with the help of the analog equation method. Numerical experiments show that a good agreement is achieved between results obtained from the proposed meshless method and available analytical solutions. It is clear that the mechanical responses in FGM differ substantially from those in their homogeneous counterparts. The proper graded parameter can lead to small stress concentration and small change in the distribution of stress fields.

References

[1] H. Parkus, Thermoelasticity, Springer, New York, 1976.

[2] S.P. Timoshenko, J.N. Goodier, Theory of Elasticity, Mcgraw Hill, 1970.

[3] C.A. Brebbia, J. Dominguez, Boundary Elements: An Introductory Course, Computational Mechanics Publications, Southampton, 1992.

[4] M.H. Sadd, Elasticity: Theory, Applications, and Numerics, Academic Press, 2009.

[5] P.K. Kythe, Fundamental Solutions for Differential Operators and Applications, Birkhauser, Boston, 1996.

[6] L. Marin, D. Lesnic, The method of fundamental solutions for the cauchy problem in two-dimensional linear elasticity, International Journal of Solids and Structures 41 (2004) 3425−3438.

[7] C.Y. Lee, H. Wang, Q.H. Qin, Method of fundamental solutions for 3D elasticity with body forces by coupling compactly supported radial basis functions, Engineering Analysis With Boundary Elements 60 (2015) 123−136.

[8] H. Wang, Q.H. Qin, Y.L. Kang, Thermoelastic meshless analysis based on method of fundamental solutions with radial basis functions approximation (in Chinese), Journal of Dalian University of Technolgy 46 (2006) 46−51.

[9] A.P.S. Selvadurai, Partial Differential Equations in Mechanics, Springer, 2000.

[10] M. Naebe, K. Shirvanimoghaddam, Functionally graded materials: a review of fabrication and properties, Applied Materials Today 5 (2016) 223−245.

[11] H. Wang, Q.H. Qin, Meshless approach for thermo-mechanical analysis of functionally graded materials, Engineering Analysis With Boundary Elements 32 (2008) 704−712.

[12] T. Tan, N. Rahbar, S.M. Allameh, S. Kwofie, D. Dissmore, K. Ghavami, W.O. Soboyejo, Mechanical properties of functionally graded hierarchical bamboo structures, Acta Biomaterialia 7 (2011) 3796−3803.

[13] E.C.N. Silva, M.C. Walters, G.H. Paulino, Modeling bamboo as a functionally graded material: lessons for the analysis of affordable materials, Journal of Materials Science 41 (2006) 6991−7004.

[14] F. Tarlochan, H. Mehboob, A. Mehboob, S.H. Chang, Influence of functionally graded pores on bone ingrowth in cementless hip prosthesis: a finite element study using mechano-regulatory algorithm, Biomechanics and Modeling in Mechanobiology 17 (2018) 701−716.

[15] M. Wehmöller, K. Neuking, M. Epple, T. Annen, H. Eufinger, Mechanical characteristics of functionally graded biodegradable implants for skull bone reconstruction, Materialwissenschaft und Werkstofftechnik 37 (2006) 413−415.

[16] A.R. Damanpack, M. Bodaghi, H. Ghassemi, M. Sayehbani, Boundary element method applied to the bending analysis of thin functionally graded plates, Latin American Journal of Solids and Structures 10 (2013) 549−570.

[17] J.T. Katsikadelis, The Boundary Element Method for Engineers and Scientists: Theory and Applications, Elsevier, 2016.

[18] B.V. Sankar, An elasticity solution for functionally graded beams, Composites Science and Technology 61 (2001) 689–696.

[19] H. Wang, Q.H. Qin, Boundary integral based graded element for elastic analysis of 2D functionally graded plates, European Journal of Mechanics – A: Solids 33 (2012) 12–23.

[20] C.O. Horgan, A.M. Chan, The pressurized hollow cylinder or disk problem for functionally graded isotropic linearly elastic materials, Journal of Elasticity 55 (1999) 43–59.

[21] M. Jabbari, S. Sohrabpour, M.R. Eslami, Mechanical and thermal stresses in a functionally graded hollow cylinder due to radially symmetric loads, International Journal of Pressure Vessels and Piping 79 (2002) 493–497.

Chapter 7

Meshless analysis for plane piezoelectric problems

Chapter outline

7.1 Introduction

The study of piezoelectricity was initiated by Jacques Curie and Pierre Curie in 1880 [1]. They found that certain crystalline materials generate an electric charge proportional to a mechanical stress. Since then new theories and applications of the field have been constantly advanced [2–9]. In particular, piezoelectric materials have been widely used in engineering as core components of transducers, sensors, and actuators for smart control [10]. The quantitative analysis of piezoelectric materials, therefore, becomes more and more important in designing these smart structures and systems. It is difficult, however, to obtain analytical solutions of coupled electroelastic fields of these structures because of the complexity of coupled mechanical-electro governing equations [11]. Consequently, approximated solutions obtained by numerical methods become a good choice for engineering analysis. At present, there have been increasing efforts in modeling electroelastic interaction in piezoelectric media using the finite element method [12–17] and the boundary element method (BEM) [18–21].

In this chapter, the boundary-type meshless method, or the method of fundamental solutions, is employed for the analysis of transversely isotropic plane piezoelectricity [22–24]. As the basic governing equations for

Methods of Fundamental Solutions in Solid Mechanics. https://doi.org/10.1016/B978-0-12-818283-3.00007-5
© 2019 Higher Education Press. Published by Elsevier Inc. All rights reserved.   **211**

two-dimensional piezoelectric problems have been provided in Chapter 2, we concentrate only on the solution procedure of the method of fundamental solutions (MFS) for such electric-elastic problems in the most part of this chapter.

7.2 Fundamental solutions for plane piezoelectricity

To obtain the fundamental solutions of plane piezoelectricity, the governing equations in terms of the primary variable $\mathbf{u} = \{u \quad w\}^{\mathrm{T}}$ and φ are recalled first. In the absence of the body force and charge density, Eqs. (2.114) and (2.115) are rewritten as

$$\mathbf{D}\begin{Bmatrix} u \\ w \\ \varphi \end{Bmatrix} = 0 \tag{7.1}$$

where $\mathbf{D}$ is the differential operator matrix

$$\mathbf{D} = \begin{bmatrix} c_{11}\dfrac{\partial^2}{\partial x^2} + c_{44}\dfrac{\partial^2}{\partial z^2} & (c_{13} + c_{44})\dfrac{\partial^2}{\partial x \partial z} & (e_{15} + e_{31})\dfrac{\partial^2}{\partial x \partial z} \\[12pt] (c_{13} + c_{44})\dfrac{\partial^2}{\partial x \partial z} & c_{44}\dfrac{\partial^2}{\partial x^2} + c_{33}\dfrac{\partial^2}{\partial z^2} & e_{15}\dfrac{\partial^2}{\partial x^2} + e_{33}\dfrac{\partial^2}{\partial z^2} \\[12pt] -(e_{15} + e_{31})\dfrac{\partial^2}{\partial x \partial z} & -\left(e_{15}\dfrac{\partial^2}{\partial x^2} + e_{33}\dfrac{\partial^2}{\partial z^2}\right) & \lambda_{11}^{\varepsilon}\dfrac{\partial^2}{\partial x^2} + \lambda_{33}^{\varepsilon}\dfrac{\partial^2}{\partial z^2} \end{bmatrix} \tag{7.2}$$

Then, the determinant of the matrix $\mathbf{D}$ is obtained as

$$|\mathbf{D}| = a\frac{\partial^6}{\partial z^6} + b\frac{\partial^6}{\partial z^4 \partial x^2} + b\frac{\partial^6}{\partial z^2 \partial x^4} + d\frac{\partial^6}{\partial x^6} \tag{7.3}$$

where

$$a = c_{44}\left(e_{33}^2 + c_{33}\lambda_{33}^{\varepsilon}\right)$$

$$b = c_{33}\left[c_{44}\lambda_{11}^{\varepsilon} + (e_{15} + e_{31})^2\right] + \lambda_{33}^{\varepsilon}\left[c_{11}c_{33} + c_{44}^2 - (c_{13} + c_{44})^2\right]$$

$$\qquad + e_{33}[2c_{44}e_{15} + c_{11}e_{33} - 2(c_{13} + c_{44})(e_{15} + e_{31})]$$

$$c = c_{44}\left[c_{11}\lambda_{33}^{\varepsilon} + (e_{15} + e_{31})^2\right] + \lambda_{11}^{\varepsilon}\left[c_{11}c_{33} + c_{44}^2 - (c_{13} + c_{44})^2\right]$$

$$\qquad + e_{15}[2c_{11}e_{33} + c_{44}e_{15} - 2(c_{13} + c_{44})(e_{15} + e_{31})]$$

$$d = c_{11}\left(e_{15}^2 + c_{44}\lambda_{11}^{\varepsilon}\right)$$

$$\tag{7.4}$$

The fundamental solutions of displacement and electric potential for the two-dimensional piezoelectric media employed in this work can be derived by means of the generalized Almansi's theorem [19], the state equations in the Fourier transformation [25], the complex function method [26], and the generalized potential functions [27]. Then the corresponding stress and electric displacement components can be obtained by substituting the displacement and electric potential components into the basic governing equations of linear piezoelectric theory. For an infinite plane piezoelectric medium under plane strain state, it is supposed that at an arbitrary source point (ξ, η), there are unit loads along x- and z-direction and an unit charge, as shown in Fig. 7.1. Then, the expressions of the displacements, electric potential, stresses, and electric displacement components caused at ay field point (x, z) can be derived as follows [25].

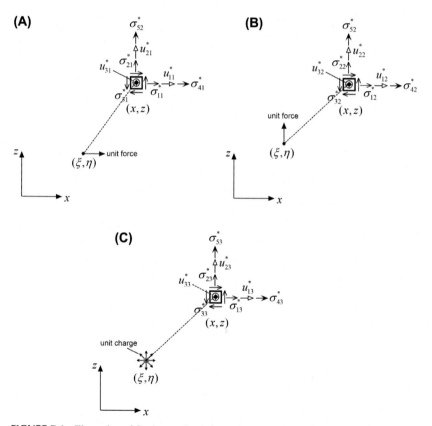

FIGURE 7.1 Illustration of fundamental solutions of plane piezoelectricity caused by (A) unit force along x-direction, (B) unit force along z-direction, and (C) unit charge.

1. The displacement and electric potential components

$$
\left\{
\begin{aligned}
u^*_{11} &= \frac{1}{\pi De11} \sum_{j=1}^{3} s_{j1} t^1_{(2j)1} \ln r_j \\[2mm]
u^*_{12} &= \frac{1}{\pi De12} \sum_{j=1}^{3} s_{j2} t^1_{(2j)2} \text{arctg} \frac{x-\xi}{s_j(z-\eta)} \\[2mm]
u^*_{13} &= -\frac{1}{\pi De13} \sum_{j=1}^{3} s_{j3} t^1_{(2j)3} \text{arctg} \frac{x-\xi}{s_j(z-\eta)}
\end{aligned}
\right.
\tag{7.5}
$$

$$
\left\{
\begin{aligned}
u^*_{21} &= \frac{1}{\pi De11} \sum_{j=1}^{3} d_{j1} t^1_{(2j)1} \text{arctg} \frac{x-\xi}{s_j(z-\eta)} \\[2mm]
u^*_{22} &= \frac{1}{\pi De12} \sum_{j=1}^{3} d_{j2} t^1_{(2j)2} \ln r_j \\[2mm]
u^*_{23} &= -\frac{1}{\pi De13} \sum_{j=1}^{3} d_{j3} t^1_{(2j)3} \ln r_j
\end{aligned}
\right.
\tag{7.6}
$$

$$
\left\{
\begin{aligned}
u^*_{31} &= \frac{1}{\pi De11} \sum_{j=1}^{3} g_{j1} t^1_{(2j)1} \text{arctg} \frac{x-\xi}{s_j(z-\eta)} \\[2mm]
u^*_{32} &= \frac{1}{\pi De12} \sum_{j=1}^{3} g_{j2} t^1_{(2j)2} \ln r_j \\[2mm]
u^*_{33} &= -\frac{1}{\pi De13} \sum_{j=1}^{3} g_{j3} t^1_{(2j)3} \ln r_j
\end{aligned}
\right.
\tag{7.7}
$$

2. The stress and electric displacement components

$$
\left\{
\begin{aligned}
\sigma^*_{11} &= \frac{1}{\pi De11} \sum_{j=1}^{3} \left[(c_{11}s_{j1} - c_{13}d_{j1}s_j - e_{31}g_{j1}s_j) t^1_{(2j)1} \frac{x-\xi}{r_j^2} \right] \\[2mm]
\sigma^*_{12} &= \frac{1}{\pi De12} \sum_{j=1}^{3} \left[(c_{11}s_{j2} + c_{13}d_{j2}s_j + e_{31}g_{j2}s_j) t^1_{(2j)2} \frac{s_j(z-\eta)}{r_j^2} \right] \\[2mm]
\sigma^*_{13} &= -\frac{1}{\pi De13} \sum_{j=1}^{3} \left[(c_{11}s_{j3} + c_{13}d_{j3}s_j + e_{31}g_{j3}s_j) t^1_{(2j)3} \frac{s_j(z-\eta)}{r_j^2} \right]
\end{aligned}
\right.
\tag{7.8}
$$

$$\begin{cases} \sigma_{21}^* = \dfrac{1}{\pi De11} \sum_{j=1}^{3} \left[\left(c_{13}s_{j1} - c_{33}d_{j1}s_j - e_{33}g_{j1}s_j \right) t_{(2j)1}^1 \dfrac{x-\xi}{r_j^2} \right] \\[2ex] \sigma_{22}^* = \dfrac{1}{\pi De12} \sum_{j=1}^{3} \left[\left(c_{13}s_{j2} + c_{33}d_{j2}s_j + e_{33}g_{j2}s_j \right) t_{(2j)2}^1 \dfrac{s_j(z-\eta)}{r_j^2} \right] \\[2ex] \sigma_{23}^* = -\dfrac{1}{\pi De13} \sum_{j=1}^{3} \left[\left(c_{13}s_{j3} + c_{33}d_{j3}s_j + e_{33}g_{j3}s_j \right) t_{(2j)3}^1 \dfrac{s_j(z-\eta)}{r_j^2} \right] \end{cases} \qquad (7.9)$$

$$\begin{cases} \sigma_{31}^* = \dfrac{1}{\pi De11} \sum_{j=1}^{3} \left[\left(c_{44}s_{j1}s_j + c_{44}d_{j1} + e_{15}g_{j1} \right) t_{(2j)1}^1 \dfrac{s_j(z-\eta)}{r_j^2} \right] \\[2ex] \sigma_{32}^* = \dfrac{1}{\pi De12} \sum_{j=1}^{3} \left[\left(-c_{44}s_{j2}s_j + c_{44}d_{j2} + e_{15}g_{j2} \right) t_{(2j)2}^1 \dfrac{x-\xi}{r_j^2} \right] \\[2ex] \sigma_{33}^* = -\dfrac{1}{\pi De13} \sum_{j=1}^{3} \left[\left(-c_{44}s_{j3}s_j + c_{44}d_{j3} + e_{15}g_{j3} \right) t_{(2j)3}^1 \dfrac{x-\xi}{r_j^2} \right] \end{cases} \qquad (7.10)$$

$$\begin{cases} \sigma_{41}^* = \dfrac{1}{\pi De11} \sum_{j=1}^{3} \left[\left(e_{15}s_{j1}s_j + e_{15}d_{j1} - \lambda_{11}g_{j1} \right) t_{(2j)1}^1 \dfrac{s_j(z-\eta)}{r_j^2} \right] \\[2ex] \sigma_{42}^* = \dfrac{1}{\pi De12} \sum_{j=1}^{3} \left[\left(-e_{15}s_{j2}s_j + e_{15}d_{j2} - \lambda_{11}g_{j2} \right) t_{(2j)2}^1 \dfrac{x-\xi}{r_j^2} \right] \\[2ex] \sigma_{43}^* = -\dfrac{1}{\pi De13} \sum_{j=1}^{3} \left[\left(-e_{15}s_{j3}s_j + e_{15}d_{j3} - \lambda_{11}g_{j3} \right) t_{(2j)3}^1 \dfrac{x-\xi}{r_j^2} \right] \end{cases} \qquad (7.11)$$

$$\begin{cases} \sigma_{51}^* = \dfrac{1}{\pi De11} \sum_{j=1}^{3} \left[\left(e_{31}s_{j1} - e_{33}d_{j1}s_j + \lambda_{33}g_{j1}s_j \right) t_{(2j)1}^1 \dfrac{x-\xi}{r_j^2} \right] \\[2ex] \sigma_{52}^* = \dfrac{1}{\pi De12} \sum_{j=1}^{3} \left[\left(e_{31}s_{j2} + e_{33}d_{j2}s_j - \lambda_{33}g_{j2}s_j \right) t_{(2j)2}^1 \dfrac{s_j(z-\eta)}{r_j^2} \right] \\[2ex] \sigma_{53}^* = -\dfrac{1}{\pi De13} \sum_{j=1}^{3} \left[\left(e_{31}s_{j3} + e_{33}d_{j3}s_j - \lambda_{33}g_{j3}s_j \right) t_{(2j)3}^1 \dfrac{s_j(z-\eta)}{r_j^2} \right] \end{cases} \qquad (7.12)$$

where the fundamental solutions u_{ij}^* ($i = 1, 2, 3; j = 1, 2, 3$) represent the induced i-directional displacement ($i = 1, 2$) or electric potential ($i = 3$) at the field point (x, z) when a unit point force along the j-direction ($j = 1, 2$) or a unit point charge ($j = 3$) is applied at the source point (ξ, η), while the fundamental solutions σ_{ij}^* ($i = 1, 2, 3, 4, 5; j = 1, 2, 3$) represent the induced stress component ($i = 1, 2, 3$) or electric displacement ($i = 4, 5$) at the field point (x, z) due to a unit point force along the j-direction ($j = 1, 2$) or a unit point charge ($j = 3$) applied at the source point (ξ, η). Here the concise notation is used for the sake of simplicity, that is,

$$u_1 = u, u_2 = w, u_3 = \varphi \tag{7.13}$$

$$\sigma_1 = \sigma_x, \sigma_2 = \sigma_z, \sigma_3 = \tau_{xz}, \sigma_4 = D_x, \sigma_5 = D_z \tag{7.14}$$

Besides, in the expressions of fundamental solutions,

$$r_j = \sqrt{(x - \xi)^2 + s_j^2(z - \eta)^2} \tag{7.15}$$

and s_i ($i = 1, 2, 3$) are the three roots of the following algebraic equation:

$$as_i^6 - bs_i^4 + cs_i^2 - d = 0 \tag{7.16}$$

Here, it is assumed $\text{Re}(s_i) > 0$ in practice.

It is worth noting that the preceding expressions of fundamental solutions were derived only for the case $s_1 \neq s_2 \neq s_3$, which are true for common piezoelectric materials like (lead zirconate titanate) PZT-4 and PZT-5H [28], and for the other two cases, i.e., $s_1 \neq s_2 = s_3$ and $s_1 = s_2 = s_3$, the related fundamental solutions can be found in the literature [25].

Additionally, the parameters included in the fundamental solutions of plane piezoelectricity are related to the piezoelectric material constants and are given next.

$$h_1 = (c_{13} + c_{44})\varepsilon_{11} + (e_{15} + e_{31})e_{15}, \quad h_2 = (c_{13} + c_{44})\varepsilon_{13} + (e_{15} + e_{31})e_{33}$$

$$h_3 = c_{11}\varepsilon_{33} + c_{44}\varepsilon_{11} + (e_{15} + e_{31})^2, \quad h_4 = c_{11}e_{33} - c_{13}(e_{15} + e_{31}) - c_{44}e_{31}$$

$$h_5 = c_{11}(c_{33}\varepsilon_{11} + e_{15}e_{33}) - c_{13}h_1, \quad h_6 = c_{13}h_2 - c_{33}h_3 - e_{33}h_4,$$
$$h_7 = c_{44}\left(c_{33}\varepsilon_{33} + e_{33}^2\right)$$

$$h_8 = c_{11}\left(c_{44}\varepsilon_{11} + e_{15}^2\right), \quad h_9 = c_{44}(h_1 - h_3) - e_{15}h_4,$$
$$h_{10} = c_{44}(h_2 - c_{44}\varepsilon_{33} - e_{15}e_{33})$$

$$h_{11} = c_{11}(\varepsilon_{11}e_{33} - \varepsilon_{33}e_{15}) - e_{31}h_1, \quad h_{12} = e_{31}h_2 - e_{33}h_3 + \varepsilon_{33}h_4$$

$$m_1 = c_{44}\varepsilon_{11} + e_{15}^2, \quad m_2 = c_{44}\varepsilon_{33} + c_{33}\varepsilon_{11} + 2e_{15}e_{33}, \quad m_3 = c_{33}\varepsilon_{33} + e_{33}^2$$

$$m_4 = c_{44}e_{31} - c_{13}e_{15}, \quad m_5 = -(c_{13} + c_{44})e_{33} + (e_{15} + e_{31})c_{33},$$
$$m_6 = c_{13}m_2 - c_{33}h_1 + e_{33}m_4$$

$$m_7 = -c_{13}m_3 + c_{33}h_2 - e_{33}m_5, \quad m_8 = c_{44}m_1 - c_{44}h_1 + e_{15}m_4,$$
$$m_9 = -c_{44}m_2 + c_{44}h_2 - e_{15}m_5$$

$$m_{10} = e_{31}m_2 - e_{33}h_1 - \varepsilon_{33}m_4, \quad m_{11} = -e_{31}m_3 + e_{33}h_2 + \varepsilon_{33}m_5$$

$$n_1 = c_{11}c_{33} - c_{13}^2 - 2c_{13}c_{44}, \quad n_2 = -c_{13}m_4 + c_{11}(c_{44}e_{33} - c_{33}e_{15}),$$
$$n_3 = c_{13}m_5 + c_{33}h_4 - e_{33}n_1$$

$$n_4 = c_{44}m_4 + c_{44}h_4 - e_{15}n_1, \quad n_5 = c_{44}(m_5 + c_{44}e_{33} - c_{33}e_{15})$$

$$n_6 = -e_{33}m_4 + c_{11}(e_{15}e_{33} + c_{44}\varepsilon_{33})$$

$$n_7 = e_{31}m_5 + e_{33}h_4 + \varepsilon_{33}n_1, \quad n_8 = -c_{44}e_{33}^2 - c_{33}c_{44}\varepsilon_{33}$$

$$a_{i1} = -c_{13}m_1 + m_6s_i^2 + m_7s_i^4, \quad b_{i1} = m_8s_i + m_9s_i^3 + c_{44}m_3s_i^5,$$
$$p_{i1} = -e_{31}m_1 + m_{10}s_i^2 + m_{11}s_i^4$$

$$d_{i1} = -h_1s_i + h_2s_i^3 \quad s_{i1} = m_1 - m_2s_i^2 + m_3s_i^4, \quad g_{i1} = m_4s_i - m_5s_i^3$$

$$n_{131} = 2m_6s_3 + 4m_7s_3^3, \quad n_{231} = m_8 + 3m_9s_3^2 + 5c_{44}m_3s_3^4,$$
$$n_{331} = 2m_{10}s_3 + 4m_{11}s_3^3$$

$$n_{431} = -h_1 + 3h_2s_3^2, \quad n_{531} = -2m_2s_3 + 4m_3s_3^3, \quad n_{631} = m_4 - 3m_5s_3^2$$

$$l_{151} = 2m_6 + 12m_7s_1^2, \quad l_{251} = 6m_9s_1 + 20c_{44}m_3s_1^3, \quad l_{351} = 2m_{10} + 12m_{11}s_1^2,$$
$$l_{451} = 6h_2s_1$$

$$l_{551} = -2m_2 + 12m_3s_1^2, \quad l_{651} = -6m_5s_1$$

$$De11 = 2(b_{31}d_{21}g_{11} - b_{21}d_{31}g_{11} - b_{31}d_{11}g_{21} + b_{11}d_{31}g_{21}$$
$$+ b_{21}d_{11}g_{31} - b_{11}d_{21}g_{31})$$

$$t_{21}^1 = -d_{31}g_{21} + d_{21}g_{31}, \quad t_{41}^1 = d_{31}g_{11} - d_{11}g_{31}, \quad t_{61}^1 = -d_{21}g_{11} + d_{11}g_{21}$$

$$De21 = 2(-d_{31}g_{11}n_{231} + d_{11}g_{31}n_{231} + b_{31}g_{11}n_{431} - b_{11}g_{31}n_{431}$$
$$- b_{31}d_{11}n_{631} + b_{11}d_{31}n_{631})$$

$$t_{21}^2 = g_{31}n_{431} - d_{31}n_{631}, \quad t_{41}^2 = -t_{41}^1, \quad t_{61}^2 = -g_{11}n_{431} + d_{11}n_{631}$$

$$De31 = 2(-g_{11}l_{451}n_{231} + d_{11}l_{651}n_{231} + g_{11}l_{251}n_{431} - b_{11}l_{651}n_{431}$$
$$- d_{11}l_{251}n_{631} + b_{11}l_{451}n_{631})$$

$$t_{21}^3 = l_{651}n_{431} - l_{451}n_{631}, \quad t_{41}^3 = -g_{11}l_{451} + d_{11}l_{651}, \quad t_{61}^3 = t_{61}^2$$

$$a_{i2} = h_7 s_i^5 + h_6 s_i^3 + h_5 s_i, \quad b_{i2} = h_{10} s_i^4 - h_9 s_i^2 - h_8, \quad p_{i2} = h_{12} s_i^3 + h_{11} s_i$$

$$d_{i2} = c_{44} e_{33} s_i^4 - h_3 s_i^2 + c_{11} e_{11}, \quad s_{i2} = h_2 s_i^3 - h_1 s_i, \quad g_{i2} = c_{44} e_{33} s_i^4 - h_4 s_i^2 + c_{11} e_{15}$$

$$n_{132} = h_5 + 3h_6 s_3^2 + 5h_7 s_3^4, \quad n_{232} = -2h_9 s_3 + 4h_{10} s_3^3, \quad n_{332} = h_{11} + 3h_{12} s_3^2$$

$$n_{432} = -2h_3 s_3 + 4c_{44} e_{33} s_3^3, \quad n_{532} = -h_1 + 3h_2 s_3^2, \quad n_{632} = -2h_4 s_3 + 4c_{44} e_{33} s_3^3$$

$$l_{152} = 6h_6 s_1 + 20h_7 s_1^3, \quad l_{252} = -2h_9 + 12h_{10} s_1^2, \quad l_{352} = 6h_{12} s_1,$$
$$l_{452} = -2h_3 + 12c_{44} e_{33} s_1^2$$

$$l_{552} = 6h_2 s_1, \quad l_{652} = -2h_4 + 12c_{44} e_{33} s_1^2$$

$$De12 = 2(-a_{32} p_{22} s_{12} + a_{22} p_{32} s_{12} + a_{32} p_{12} s_{22} - a_{12} p_{32} s_{22}$$
$$- a_{22} p_{12} s_{32} + a_{12} p_{22} s_{32})$$

$$t_{22}^1 = p_{32} s_{22} - p_{22} s_{32}, \quad t_{42}^1 = -p_{32} s_{12} + p_{12} s_{32}, \quad t_{62}^1 = p_{22} s_{12} - p_{12} s_{22}$$

$$De22 = 2(-p_{32} s_{12} n_{132} + p_{12} s_{32} n_{132} + a_{32} s_{12} n_{332} - a_{12} s_{32} n_{332}$$
$$- a_{32} p_{12} n_{532} + a_{12} p_{32} n_{532})$$

$$t_{22}^2 = s_{32} n_{332} - p_{32} n_{532}, \quad t_{42}^2 = -p_{32} s_{12} + p_{12} s_{32}, \quad t_{62}^2 = -s_{12} n_{332} + p_{12} n_{532}$$

$$De32 = 2(-s_{12} l_{352} n_{132} + p_{12} l_{552} n_{132} + s_{12} l_{152} n_{332} - a_{12} l_{552} n_{332}$$
$$- p_{12} l_{152} n_{532} + a_{12} l_{352} n_{532})$$

$$t_{22}^3 = l_{552} n_{332} - l_{352} n_{532}, \quad t_{42}^3 = -s_{12} l_{352} + p_{12} l_{552}, \quad t_{62}^3 = -s_{12} n_{332} + p_{12} n_{532}$$

$$a_{i3} = n_2 s_i + n_3 s_i^3, \quad b_{i3} = -n_4 s_i^2 + n_5 s_i^4, \quad p_{i3} = n_6 s_i + n_7 s_i^3 + n_8 s_i^5$$

$$d_{i3} = -c_{11} e_{15} + h_4 s_i^2 - c_{44} e_{33} s_i^4, \quad s_{i3} = m_5 s_i^3 - m_4 s_i,$$
$$g_{i3} = c_{33} c_{44} s_i^4 - n_1 s_i^2 + c_{11} c_{44}$$

$$n_{133} = n_2 + 3n_3 s_3^2, \quad n_{233} = -2n_4 s_3 + 4n_5 s_3^3, \quad n_{333} = n_6 + 3n_7 s_3^2 + 5n_8 s_3^4$$

$$n_{433} = 2h_4 s_3 - 4c_{44} e_{33} s_3^3, \quad n_{533} = -m_4 + 3m_5 s_3^2, \quad n_{633} = -2n_1 s_3 + 4c_{33} c_{44} s_3^3$$

$$l_{153} = 6n_3 s_1, \quad l_{253} = -2n_4 + 12n_5 s_1^2, \quad l_{353} = 6n_7 s_1 + 20n_8 s_1^3,$$
$$l_{453} = 2h_4 - 12c_{44} e_{33} s_1^2$$

$$l_{553} = 6m_5 s_1, \quad l_{653} = -2n_1 + 12c_{33} s_1^2$$

$$De13 = 2(-a_{33}p_{23}s_{13} + a_{23}p_{33}s_{13} + a_{33}p_{13}s_{23} - a_{13}p_{33}s_{23}$$
$$- a_{23}p_{13}s_{33} + a_{13}p_{23}s_{33})$$

$$t_{23}^1 = -a_{33}s_{23} + a_{23}s_{33}, \quad t_{43}^1 = a_{33}s_{13} - a_{13}s_{33}, \quad t_{63}^1 = -a_{23}s_{13} + a_{13}s_{23}$$

$$De23 = 2(-p_{33}s_{13}n_{133} + p_{13}s_{33}n_{133} + a_{33}s_{13}n_{333} - a_{13}s_{33}n_{333}$$
$$- a_{33}p_{13}n_{533} + a_{13}p_{33}n_{533})$$

$$t_{23}^2 = -s_{33}n_{133} + a_{33}n_{533}, \quad t_{43}^2 = a_{33}s_{13} - a_{13}s_{33}, \quad t_{63}^2 = s_{13}n_{133} - a_{13}n_{533}$$

$$De33 = 2(-s_{13}l_{353}n_{133} + p_{13}l_{553}n_{133} + s_{13}l_{153}n_{333} - a_{13}l_{553}n_{333}$$
$$- p_{13}l_{153}n_{533} + a_{13}l_{353}n_{533})$$

$$t_{23}^3 = -l_{553}n_{133} + l_{155}n_{533}, \quad t_{43}^3 = s_{13}l_{153} - a_{13}l_{553}, \quad t_{63}^3 = s_{13}n_{133} - a_{13}n_{533}$$

7.3 Solution procedure for plane piezoelectricity

Following the basic idea of the MFS, a few of the fictitious source points are placed outside the solution domain Ω, as shown in Fig. 7.2. It is assumed that the total numbers of source points are N_s.

In the case of absence of body forces and charge density, the displacement and electric potential of the homogeneous governing Eq. (7.1) can be approximated by

$$u_i(\mathbf{x}) = \sum_{k=1}^{N_s} \left[u_{i1}^*(\mathbf{x}, \mathbf{y}_k)\alpha_{k1} + u_{i2}^*(\mathbf{x}, \mathbf{y}_k)\alpha_{k2} + u_{i3}^*(\mathbf{x}, \mathbf{y}_k)\alpha_{k3} \right], \quad i = 1, 2, 3$$

(7.17)

where α_{kj} ($k = 1 \to N_s, j = 1, 2, 3$) denote the point force ($j = 1, 2$) or point charge ($j = 3$) at the source point $\mathbf{y}_k$.

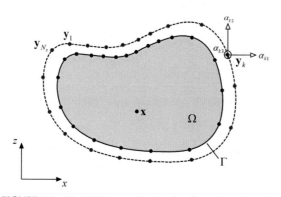

FIGURE 7.2 The MFS approximation for plane piezoelectricity.

In matrix form, Eq. (7.17) can be rewritten as

$$
\begin{Bmatrix} u_1(\mathbf{x}) \\ u_2(\mathbf{x}) \\ u_3(\mathbf{x}) \end{Bmatrix} = \begin{Bmatrix} u(\mathbf{x}) \\ w(\mathbf{x}) \\ \varphi(\mathbf{x}) \end{Bmatrix} = \begin{bmatrix} \mathbf{U}_{11}^*(\mathbf{x}) & \mathbf{U}_{12}^*(\mathbf{x}) & \cdots & \mathbf{U}_{1N_s}^*(\mathbf{x}) \\ \mathbf{U}_{21}^*(\mathbf{x}) & \mathbf{U}_{22}^*(\mathbf{x}) & \cdots & \mathbf{U}_{2N_s}^*(\mathbf{x}) \\ \mathbf{U}_{31}^*(\mathbf{x}) & \mathbf{U}_{32}^*(\mathbf{x}) & \cdots & \mathbf{U}_{3N_s}^*(\mathbf{x}) \end{bmatrix} \begin{Bmatrix} \alpha_1 \\ \alpha_2 \\ \vdots \\ \alpha_{N_s} \end{Bmatrix}
$$

$$(7.18)$$

where

$$
\mathbf{U}_{ik}^*(\mathbf{x}) = \begin{bmatrix} u_{i1}^*(\mathbf{x}, \mathbf{y}_k) & u_{i2}^*(\mathbf{x}, \mathbf{y}_k) & u_{i3}^*(\mathbf{x}, \mathbf{y}_k) \end{bmatrix} \tag{7.19}
$$

$$
\boldsymbol{\alpha}_k = \begin{bmatrix} \alpha_{k1} & \alpha_{k2} & \alpha_{k3} \end{bmatrix}^{\mathrm{T}} \tag{7.20}
$$

Similarly, the stress components and electric displacements at arbitrary point $\mathbf{x} \in \Omega$ can be approximated by the linear combination of related fundamental solutions as

$$
\sigma_i(\mathbf{x}) = \sum_{k=1}^{N_s} \left[\sigma_{i1}^*(\mathbf{x}, \mathbf{y}_k)\alpha_{k1} + \sigma_{i2}^*(\mathbf{x}, \mathbf{y}_k)\alpha_{k2} + \sigma_{i3}^*(\mathbf{x}, \mathbf{y}_k)\alpha_{k3} \right], \quad i = 1, 2, 3, 4, 5
$$

$$(7.21)$$

or in matrix form

$$
\begin{Bmatrix} \sigma_1(\mathbf{x}) \\ \sigma_2(\mathbf{x}) \\ \sigma_3(\mathbf{x}) \\ \sigma_4(\mathbf{x}) \\ \sigma_5(\mathbf{x}) \end{Bmatrix} = \begin{Bmatrix} \sigma_x(\mathbf{x}) \\ \sigma_z(\mathbf{x}) \\ \tau_{xz}(\mathbf{x}) \\ D_x(\mathbf{x}) \\ D_z(\mathbf{x}) \end{Bmatrix} = \begin{bmatrix} \mathbf{S}_{11}^*(\mathbf{x}) & \mathbf{S}_{12}^*(\mathbf{x}) & \cdots & \mathbf{S}_{1N_s}^*(\mathbf{x}) \\ \mathbf{S}_{21}^*(\mathbf{x}) & \mathbf{S}_{22}^*(\mathbf{x}) & \cdots & \mathbf{S}_{2N_s}^*(\mathbf{x}) \\ \mathbf{S}_{31}^*(\mathbf{x}) & \mathbf{S}_{32}^*(\mathbf{x}) & \cdots & \mathbf{S}_{3N_s}^*(\mathbf{x}) \\ \mathbf{S}_{41}^*(\mathbf{x}) & \mathbf{S}_{42}^*(\mathbf{x}) & \cdots & \mathbf{S}_{4N_s}^*(\mathbf{x}) \\ \mathbf{S}_{51}^*(\mathbf{x}) & \mathbf{S}_{52}^*(\mathbf{x}) & \cdots & \mathbf{S}_{5N_s}^*(\mathbf{x}) \end{bmatrix} \begin{Bmatrix} \alpha_1 \\ \alpha_2 \\ \vdots \\ \alpha_{N_s} \end{Bmatrix}
$$

$$(7.22)$$

where

$$
\mathbf{S}_{ik}^*(\mathbf{x}) = \begin{bmatrix} \sigma_{i1}^*(\mathbf{x}, \mathbf{y}_k) & \sigma_{i2}^*(\mathbf{x}, \mathbf{y}_k) & \sigma_{i3}^*(\mathbf{x}, \mathbf{y}_k) \end{bmatrix} \tag{7.23}
$$

Further, the boundary traction forces at the boundary point can be determined. For the sake of convenience, based on Eqs. (2.102)–(2.104), the boundary tractions and normal electric displacement at the boundary point $\mathbf{x} \in \Gamma$ are written in matrix form

$$
\begin{Bmatrix} t_x(\mathbf{x}) \\ t_z(\mathbf{x}) \end{Bmatrix} = \begin{bmatrix} n_x & 0 & n_z \\ 0 & n_z & n_x \end{bmatrix} \begin{Bmatrix} \sigma_x(\mathbf{x}) \\ \sigma_z(\mathbf{x}) \\ \tau_{xz}(\mathbf{x}) \end{Bmatrix}, \quad \mathbf{x} \in \Gamma \tag{7.24}
$$

$$
D_n(\mathbf{x}) = \begin{bmatrix} n_x & n_z \end{bmatrix} \begin{Bmatrix} D_x(\mathbf{x}) \\ D_z(\mathbf{x}) \end{Bmatrix}, \quad \mathbf{x} \in \Gamma \tag{7.25}
$$

where n_x and n_z are components of unit normal vector at the boundary point $\mathbf{x}$.

Next, combining Eqs. (7.24) and (7.25) gives the following more concise form:

$$
\left\{
\begin{array}{c}
t_x(\mathbf{x}) \\
t_z(\mathbf{x}) \\
D_n(\mathbf{x})
\end{array}
\right\}
=
\begin{bmatrix}
n_x & 0 & n_z & 0 & 0 \\
0 & n_z & n_x & 0 & 0 \\
0 & 0 & 0 & n_x & n_z
\end{bmatrix}
\left\{
\begin{array}{c}
\sigma_x \\
\sigma_z \\
\tau_{xz} \\
D_x \\
D_z
\end{array}
\right\},
\quad \mathbf{x} \in \Gamma
\qquad (7.26)
$$

The substitution of the stress approximations (7.22) into Eq. (7.26) gives

$$
\left\{
\begin{array}{c}
t_x(\mathbf{x}) \\
t_z(\mathbf{x}) \\
D_n(\mathbf{x})
\end{array}
\right\}
=
\begin{bmatrix}
\mathbf{T}_{11}^*(\mathbf{x}) & \mathbf{T}_{12}^*(\mathbf{x}) & \cdots & \mathbf{T}_{1N_s}^*(\mathbf{x}) \\
\mathbf{T}_{21}^*(\mathbf{x}) & \mathbf{T}_{22}^*(\mathbf{x}) & \cdots & \mathbf{T}_{2N_s}^*(\mathbf{x}) \\
\mathbf{T}_{31}^*(\mathbf{x}) & \mathbf{T}_{32}^*(\mathbf{x}) & \cdots & \mathbf{T}_{3N_s}^*(\mathbf{x})
\end{bmatrix}
\left\{
\begin{array}{c}
\alpha_1 \\
\alpha_2 \\
\vdots \\
\alpha_{N_s}
\end{array}
\right\},
\quad \mathbf{x} \in \Gamma
\qquad (7.27)
$$

where

$$
\begin{bmatrix}
\mathbf{T}_{11}^*(\mathbf{x}) & \mathbf{T}_{12}^*(\mathbf{x}) & \cdots & \mathbf{T}_{1N_s}^*(\mathbf{x}) \\
\mathbf{T}_{21}^*(\mathbf{x}) & \mathbf{T}_{22}^*(\mathbf{x}) & \cdots & \mathbf{T}_{2N_s}^*(\mathbf{x}) \\
\mathbf{T}_{31}^*(\mathbf{x}) & \mathbf{T}_{32}^*(\mathbf{x}) & \cdots & \mathbf{T}_{3N_s}^*(\mathbf{x})
\end{bmatrix}
$$

$$
=
\begin{bmatrix}
n_x & 0 & n_z & 0 & 0 \\
0 & n_z & n_x & 0 & 0 \\
0 & 0 & 0 & n_x & n_z
\end{bmatrix}
\begin{bmatrix}
S_{11}^*(\mathbf{x}) & S_{12}^*(\mathbf{x}) & \cdots & S_{1N_s}^*(\mathbf{x}) \\
S_{21}^*(\mathbf{x}) & S_{22}^*(\mathbf{x}) & \cdots & S_{2N_s}^*(\mathbf{x}) \\
S_{31}^*(\mathbf{x}) & S_{32}^*(\mathbf{x}) & \cdots & S_{3N_s}^*(\mathbf{x}) \\
S_{41}^*(\mathbf{x}) & S_{42}^*(\mathbf{x}) & \cdots & S_{4N_s}^*(\mathbf{x}) \\
S_{51}^*(\mathbf{x}) & S_{52}^*(\mathbf{x}) & \cdots & S_{5N_s}^*(\mathbf{x})
\end{bmatrix}
\qquad (7.28)
$$

Obviously, the approximations (7.17) or (7.18) analytically satisfy the homogeneous governing (7.1). It means that the approximations (7.17) or (7.18) just need to satisfy the applied boundary conditions for determining unknown coefficients in them.

To do this, N_b collocations are chosen along the boundary of the computing domain. It is assumed that the displacement and electric potential are specified on the boundary segment of Γ_1, the traction and electric displacement are specified on the boundary segment of Γ_2, and $\Gamma_1 \cup \Gamma_2 = \Gamma$ and $\Gamma_1 \cap \Gamma_2 = \emptyset$. Therefore, by satisfaction of the approximations (7.17) and (7.21) with respect to the specific boundary conditions at the boundary collocations, we have

$$
\left\{
\begin{array}{c}
u_1(\mathbf{x}_i) \\
u_2(\mathbf{x}_i) \\
u_3(\mathbf{x}_i)
\end{array}
\right\}
=
\begin{bmatrix}
U_{11}^*(\mathbf{x}_i) & U_{12}^*(\mathbf{x}_i) & \cdots & U_{1N_s}^*(\mathbf{x}_i) \\
U_{21}^*(\mathbf{x}_i) & U_{22}^*(\mathbf{x}_i) & \cdots & U_{2N_s}^*(\mathbf{x}_i) \\
U_{31}^*(\mathbf{x}_i) & U_{32}^*(\mathbf{x}_i) & \cdots & U_{3N_s}^*(\mathbf{x}_i)
\end{bmatrix}
\left\{
\begin{array}{c}
\alpha_1 \\
\alpha_2 \\
\vdots \\
\alpha_{N_s}
\end{array}
\right\}
$$

$$
=
\left\{
\begin{array}{c}
\overline{u}(\mathbf{x}_i) \\
\overline{w}(\mathbf{x}_i) \\
\overline{\phi}(\mathbf{x}_i)
\end{array}
\right\},
\quad \mathbf{x}_i \in \Gamma_1, \quad i = 1, 2, \ldots, N_{b1}
\qquad (7.29)
$$

$$\left\{\begin{array}{c} t_x(\mathbf{x}_j) \\ t_z(\mathbf{x}_j) \\ D_n(\mathbf{x}_j) \end{array}\right\} = \left[\begin{array}{cccc} \mathbf{T}^*_{11}(\mathbf{x}_j) & \mathbf{T}^*_{12}(\mathbf{x}_j) & \cdots & \mathbf{T}^*_{1N_s}(\mathbf{x}_j) \\ \mathbf{T}^*_{21}(\mathbf{x}_j) & \mathbf{T}^*_{22}(\mathbf{x}_j) & \cdots & \mathbf{T}^*_{2N_s}(\mathbf{x}_j) \\ \mathbf{T}^*_{31}(\mathbf{x}_j) & \mathbf{T}^*_{32}(\mathbf{x}_j) & \cdots & \mathbf{T}^*_{3N_s}(\mathbf{x}_j) \end{array}\right] \left\{\begin{array}{c} \alpha_1 \\ \alpha_2 \\ \vdots \\ \alpha_{N_s} \end{array}\right\}$$

(7.30)

$$= \left\{\begin{array}{c} \bar{t}_x(\mathbf{x}_j) \\ \bar{t}_z(\mathbf{x}_j) \\ -\bar{q}(\mathbf{x}_j) \end{array}\right\}, \quad \mathbf{x}_j \in \Gamma_2, \ j = 1, 2, \ldots N_{b2}$$

where N_{b1} and N_{b2} are the number of boundary collocations on the segment Γ_1 and Γ_2, respectively, and $N_{b1} + N_{b2} = N_b$.

Eqs. (7.29) and (7.30) can be written in the matrix form as

$$\mathbf{H}_{(3N_b \times 3N_s)} \boldsymbol{\alpha}_{(3N_s \times 1)} = \mathbf{P}_{(3N_b \times 1)}$$

(7.31)

which can be solved numerically to determine all unknown coefficients. Then, the field quantities at any point in the domain can be evaluated by Eqs. (7.18) and (7.22).

7.4 Results and discussion

To verify the meshless method presented in the previous section, several numerical examples are considered, and the calculated results are compared with the available analytical solutions.

7.4.1 Simple tension of a piezoelectric prism

In the first example, a simple tension of a piezoelectric prism is taken into consideration, and the piezoelectric model under the plane strain state is shown in Fig. 7.3. The corresponding boundary conditions are as follows.

Mechanical boundary conditions:

$$t_x = 0, t_z = 0 \quad \text{along } x = \pm a$$
$$t_x = 0, t_z = p \quad \text{along } z = \pm h$$

Electric boundary conditions:

$$D_n = 0 \quad \text{along all boundaries}$$

The analytical solutions of this problem are given as [19]

$$u = \sum_{i=1}^{3} a_i x$$

$$w = -\sum_{i=1}^{3} \alpha_{i1} a_i s_i z$$

(7.32)

$$\phi = -\sum_{i=1}^{3} \alpha_{i2} a_i s_i z$$

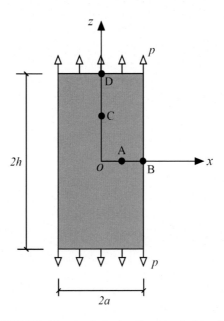

FIGURE 7.3   The simple tension of a piezoelectric prism.

where

$$a_1 = \frac{p}{D}\left(k_{11}^3 k_{13}^2 - k_{11}^2 k_{13}^3\right)$$

$$a_2 = \frac{p}{D}\left(k_{11}^1 k_{13}^3 - k_{11}^3 k_{13}^1\right), \quad D = \begin{vmatrix} k_{11}^1 & k_{11}^2 & k_{11}^3 \\ k_{12}^1 & k_{12}^2 & k_{12}^3 \\ k_{13}^1 & k_{13}^2 & k_{13}^3 \end{vmatrix} \tag{7.33}$$

$$a_3 = \frac{p}{D}\left(k_{11}^2 k_{13}^1 - k_{11}^1 k_{13}^2\right)$$

with

$$\begin{aligned} k_{11}^i &= c_{11} - c_{13}\alpha_{i1}s_i - e_{31}\alpha_{i2}s_i \\ k_{12}^i &= c_{13} - c_{33}\alpha_{i1}s_i - e_{33}\alpha_{i2}s_i, \quad i = 1,2,3 \\ k_{13}^i &= e_{31} - e_{33}\alpha_{i1}s_i + \lambda_{33}^\varepsilon \alpha_{i2}s_i \end{aligned} \tag{7.34}$$

$$\begin{aligned} \alpha_{i1} &= \frac{c_{11}\lambda_{11}^\varepsilon - m_3 s_i^2 + c_{44}\lambda_{33}^\varepsilon s_i^4}{\left(m_1 - m_2 s_i^2\right)s_i} \\ \alpha_{i2} &= \frac{c_{11}e_{15} - m_4 s_i^2 + c_{44}e_{33}s_i^4}{\left(m_1 - m_2 s_i^2\right)s_i}, \quad i = 1,2,3 \end{aligned} \tag{7.35}$$

and

$$\begin{aligned} m_1 &= (c_{13} + c_{44})\lambda_{11}^\varepsilon + (e_{15} + e_{31})e_{15} \\ m_2 &= (c_{13} + c_{44})\lambda_{33}^\varepsilon + (e_{15} + e_{31})e_{33} \\ m_3 &= c_{11}\lambda_{33}^\varepsilon + c_{44}\lambda_{11}^\varepsilon + (e_{15} + e_{31})^2 \\ m_4 &= c_{11}e_{33} - c_{13}(e_{15} + e_{31}) - c_{44}e_{31} \end{aligned} \tag{7.36}$$

TABLE 7.1 Material properties for PZT-4 piezoelectric ceramic.

Elastic constants (N/m^2)	$c_{11} = 12.6 \times 10^{10}$, $c_{12} = 7.78 \times 10^{10}$, $c_{13} = 7.43 \times 10^{10}$ $c_{33} = 11.5 \times 10^{10}$, $c_{44} = 2.56 \times 10^{10}$
Piezoelectric constants (C/m^2)	$e_{31} = -5.2$, $e_{33} = 15.1$, $e_{15} = 12.7$
Dielectric constants (C/Vm)	$\lambda_{11}^{\varepsilon} = 6.463 \times 10^{-9}$, $\lambda_{33}^{\varepsilon} = 5.622 \times 10^{-9}$

In the computation, we set $a = 3$ m, $h = 10$ m, and $P = 10$ Pa. The PZT-4 piezoelectric ceramic is taken as an example for computation. Its mechanical and electric constants are given in Table 7.1.

With 44 boundary collocations and the same number of virtual source points (see Fig. 7.4), the numerical results can be obtained by setting $\gamma = 1$. It is worth noting that the points B and D in Fig. 7.3 are constrained by setting $u_D = 0$ and $w_B = 0$ to remove the rigid mode. Some results at the four points A (2, 0), B (3, 0), C (0, 5), and D (0, 10) are given in Table 7.2. It is found from Table 7.2 that the present meshless method can achieve almost the same results as the analytical solutions. Moreover, the effects of the distance of virtual and physical boundaries to the electric potential ϕ and the stress component σ_z at point (2, 5) are investigated, and the results are given in Figs. 7.5 and 7.6, from which it is observed that the accuracy of the results will become worse if the distance of the virtual and physical boundaries is too small. This can be mainly attributed to the singular interference of fundamental solutions.

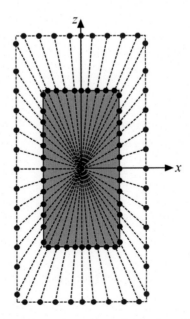

FIGURE 7.4 Collocations on the virtual and physical boundaries for the rectangular domain.

TABLE 7.2 The comparison of numerical and exact solutions (data in parenthesis).

	$u(\times 10^{-10}\,m)$	$w(\times 10^{-9}\,m)$	$\varphi(V)$	$\sigma_x(Pa)$	$\sigma_z(Pa)$	$D_z(Cm^{-2})$
A	−0.7222 (−0.7222)	0.0000 (0.0000)	0.0000 (0.0000)	0.0000 (0.0000)	9.9999 (10.000)	0.0000 (0.0000)
B	−1.0833 (−1.0834)	0.0000 (0.0000)	0.0000 (0.0000)	0.0000 (0.0000)	10.0001 (10.000)	0.0000 (0.0000)
C	0.0000 (0.0000)	0.3915 (0.3915)	1.2183 (1.2183)	0.0000 (0.0000)	10.0002 (10.000)	0.0000 (0.0000)
D	0.0000 (0.0000)	0.7829 (0.7829)	(2.4367) (2.4367)	0.0000 (0.0000)	10.000 (10.000)	0.0000 (0.0000)

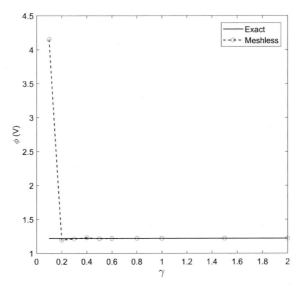

FIGURE 7.5 Effect of γ to the electric potential ϕ at point (2, 5).

7.4.2 An infinite piezoelectric plane with a circular hole under remote tension

As indicated in Fig. 7.7, an infinite piezoelectric plane with a circular hole of radius a is considered. It is assumed that the hole boundary is free of stress and electric charge, and an infinite tensile stress σ_z^{∞} along the z-direction is applied. The corresponding analytical solution can be found in Ref. [29].

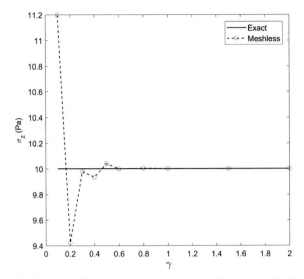

FIGURE 7.6 Effect of γ to the electric potential σ_z at point (2, 5).

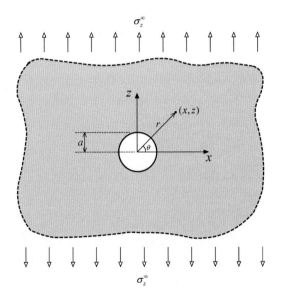

FIGURE 7.7 An infinite piezoelectric plane with a circular hole under remote tension.

In the calculation, we take $a = 1$ and $\sigma_z^\infty = 10$, and the piezoelectric material employed is the same as that in the first example. A total of 20 boundary collocations and 20 virtual source points are applied on the physical and virtual circular boundaries for the implementation of MFS, as indicated in Fig. 7.8. If the radius of the virtual boundary is $a_v = 0.7$, the parameter is $\gamma = (a_v - a)/a = -0.3$ for such a case.

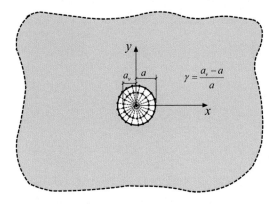

FIGURE 7.8 Collocations on the virtual and physical boundaries for infinite domain weaken with a circular hole.

Figs. 7.9 and 7.10 present distributions of the radial stress σ_r and the hoop stress σ_θ with respect to the radial spatial variable r along the line $\theta = 0$, respectively. It is seen from Fig. 7.9 that the radial stress σ_r increases rapidly from 0 to a peak value of 4.34 at about $r = 1.5$ and then decreases dramatically. Results in Fig. 7.10 indicate that the hoop stress σ_θ has maximum value on the rim of the hole, and the corresponding stress concentration coefficient is found to be 2.667 (the analytical solution is 2.721 [29]). Besides, it is shown from Figs. 7.9 and 7.10 that both σ_r and σ_θ at infinity have steady values, which correspond to the uniaxial stress state at infinity, that is, $\sigma_r = 0$ and $\sigma_\theta = \sigma_z^\infty$.

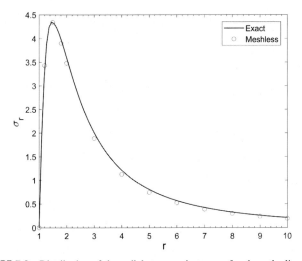

FIGURE 7.9 Distribution of the radial stress σ_r in terms of r along the line $\theta = 0$.

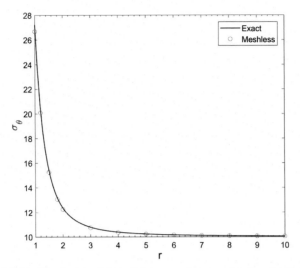

FIGURE 7.10 Distribution of the radial stress σ_θ in terms of r along the line $\theta = 0$.

Figs. 7.11 and 7.12 show the distributions of σ_r and σ_θ when the spatial variable r approaches infinity on the line $\theta = \pi/2$. It is found from Fig. 7.11 that the radial stress σ_r on the line $\theta = \pi/2$ increases gradually from zero to a steady state σ_z^∞ in terms of the spatial variable r. Meanwhile, Fig. 7.12 indicates that σ_θ is negative when r is small and then it goes to zero rapidly when r increases. This means that there is a compressive stress region over the

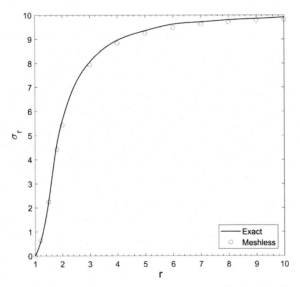

FIGURE 7.11 Distribution of the radial stress σ_r in terms of r along the line $\theta = \pi/2$.

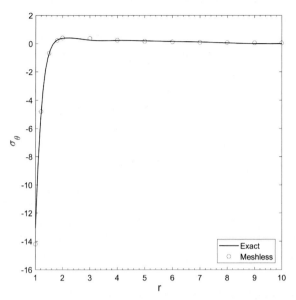

FIGURE 7.12 Distribution of the radial stress σ_θ in terms of r along the line $\theta = \pi/2$.

hole. Similarly, both σ_r and σ_θ at infinity have steady values that correspond to the uniaxial stress state at infinity.

Next, the distributions of σ_θ and D_θ along the rim of the circular hole are displayed in Figs. 7.13 and 7.14, from which it is observed that σ_θ changes from tensile stress to compressive stress and the transition point locates at about 62 degrees, while D_θ reaches a minimum value at about 65 degrees.

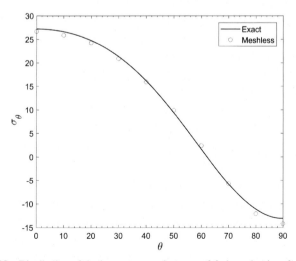

FIGURE 7.13 Distribution of the hoop stress σ_θ in terms of θ along the rim of circular hole.

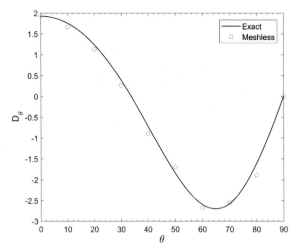

FIGURE 7.14 Distribution of the hoop electric displacement D_θ in terms of θ along the rim of circular hole.

Finally, all numerical results from the MFS show good agreement with the analytical solutions. Thus, the effectiveness and accuracy of the MFS for plane piezoelectric analysis is validated.

7.4.3 An infinite piezoelectric plane with a circular hole subject to internal pressure

In the final example, we consider the piezoelectric problem described in the literature [30], an infinite piezoelectric plane embedded with a circular cavity subject to internal pressure P as seen in Fig. 7.15. The piezoelectric material constants used in the computation are listed in Table 7.3.

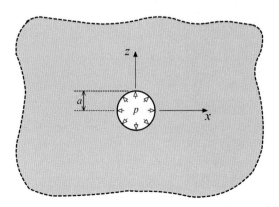

FIGURE 7.15 An infinite piezoelectric plane with a circular hole subject to internal pressure.

TABLE 7.3 Material properties for piezoelectric material [30].

Elastic constants (N/m²)	$c_{11} = 14.1 \times 10^{10}$, $c_{13} = 7.57 \times 10^{10}$ $c_{33} = 11.6 \times 10^{10}$, $c_{44} = 2.53 \times 10^{10}$
Piezoelectric constants (C/m²)	$e_{31} = -5.3$, $e_{33} = 15.5$, $e_{15} = 13.0$
Dielectric constants (C/Vm)	$\lambda_{11}^{\varepsilon} = 6.37 \times 10^{-9}$, $\lambda_{33}^{\varepsilon} = 5.53 \times 10^{-9}$

In the computation, the plane strain state is considered. The radius of the circular cavity is assumed to be 1 m, and the internal pressure is P = 10 Pa. To demonstrate the effectiveness and accuracy of the MFS, the numerical results from the BEM [30] are used for comparison. A total of 32 virtual sources are placed on a circular fictitious boundary, whose location is determined by the parameter $\gamma = -0.5$. Additionally, 64 boundary nodes are chosen from the physical boundary for enforcing the satisfaction of prescribed boundary conditions at them.

Figs. 7.16−7.18 present the variations of the primary variable u, w, and ϕ, respectively, with respect to the polar angle θ changing from 0 to 90 degrees along the rim of a circular hole. The results from Figs. 7.16−7.18 show that the displacement component u reaches its maximum value of 1.0626×10^{-11} m at the point (1, 0), the displacement component w has the maximum value of 0.9902×10^{-11} m at the point (1, 0), and the electric potential ϕ gets the

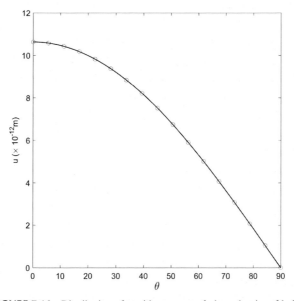

FIGURE 7.16 Distribution of u with respect to θ along the rim of hole.

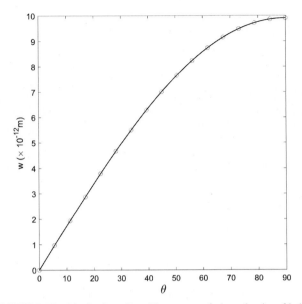

FIGURE 7.17 Distribution of w with respect to θ along the rim of hole.

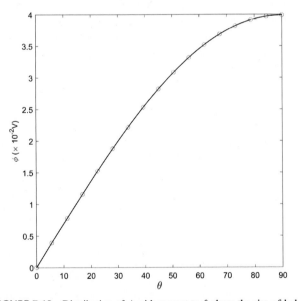

FIGURE 7.18 Distribution of ϕ with respect to θ along the rim of hole.

maximum value of 0.03,988 V at the same point $(1, 0)$. Besides, the variations of the hoop stress σ_θ and the hoop electric displacement D_θ along the circular boundary are provided in Figs. 7.19 and 7.20, respectively. It is found that when the polar angle θ is in the range of [0, 180 degrees], σ_θ and D_θ show obviously

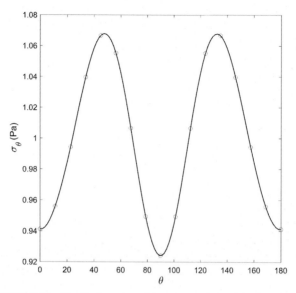

FIGURE 7.19 Distribution of σ_θ with respect to θ along the rim of hole.

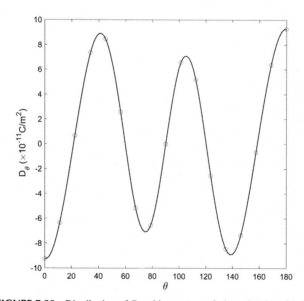

FIGURE 7.20 Distribution of D_θ with respect to θ along the rim of hole.

symmetry or antisymmetry. σ_θ achieves the maximum value 1.0666 Pa at the locations $\theta = 45$ and 135 degrees, while D_θ has the maximum and minimum values of $\pm 0.9257 \times 10^{-10}\,\mathrm{C/m^2}$ at the locations $\theta = 0$ and 180 degrees, respectively. All these extreme results agree with those from the BEM [30], as indicated in Table 7.4.

TABLE 7.4 Comparison of extreme results from the MFS and the BEM.

	u ($\times 10^{-11}$ m)	w ($\times 10^{-11}$ m)	ϕ ($\times 10^{-2}$ V)	σ_θ (Pa)	D_θ ($\times 10^{-12}$ C/m^2)
MFS	1.0626	0.9902	3.988	1.0666	9.257
BEM	1.0525	0.9808	3.919	1.067	9.35

7.5 Remarks

In this chapter, the MFS is developed for the numerical analysis of electro-mechanical interaction in plane piezoelectric materials. The approximate solutions in the MFS are assumed to be the combination of fundamental solutions of the problem. As a result, the analytical satisfaction of piezoelectric governing equations in terms of the primary variables makes that the approximate solutions in the MFS just need to satisfy the prescribed boundary conditions at boundary nodes on the physical boundary. The solution procedure of the MFS is very simple, and its numerical results indicate that the MFS can be used for the solutions of complex plane piezoelectricity effectively and flexibly.

References

[1] P. Curie, J. Curie, Dévelopment, par pression, de l'électricité polaire dans les cristaux hémièdres à faces inclinées, Comptes Rendus de l'Académie des Sciences 91 (1980) 294−295.

[2] Q.H. Qin, Thermoelectroelastic solution for elliptic inclusions and application to crack−inclusion problems, Applied Mathematical Modelling 25 (2000) 1−23.

[3] Q.H. Qin, 2D Green's functions of defective magnetoelectroelastic solids under thermal loading, Engineering Analysis with Boundary Elements 29 (2005) 577−585.

[4] Q.H. Qin, Q.S. Yang, Macro-micro Theory on Multifield Coupling Behaivor of Heterogenous Materials, Higher Education Press and Springer, Beijing, 2008.

[5] Q.H. Qin, J.Q. Ye, Thermoelectroelastic solutions for internal bone remodeling under axial and transverse loads, International Journal of Solids and Structures 41 (2004) 2447−2460.

[6] S.W. Yu, Q.H. Qin, Damage analysis of thermopiezoelectric properties: Part II. Effective crack model, Theoretical and Applied Fracture Mechanics 25 (1996) 279−288.

[7] Q.H. Qin, Green's function for thermopiezoelectric plates with holes of various shapes, Archive of Applied Mechanics 69 (1999) 406−418.

[8] Q.H. Qin, Green's functions of magnetoelectroelastic solids with a half-plane boundary or bimaterial interface, Philosophical Magazine Letters 84 (2004) 771−779.

[9] Q.H. Qin, Y.W. Mai, Crack branch in piezoelectric bimaterial system, International Journal of Engineering Science 38 (2000) 673−693.

[10] T. Ikeda, Fundamentals of Piezoelectricity, Oxford University Press, New York, 1996.

[11] W.X. Zhang, H. Wang, Axisymmetric boundary condition problems for transversely isotropic piezoelectric materials, Mechanics Research Communications 87 (2018) 7−12.

[12] H. Allik, T.J.R. Hughes, Finite element method for piezoelectric vibration, International Journal for Numerical Methods in Engineering 2 (1970) 151–157.

[13] K.Y. Lam, X.Q. Peng, G.R. Liu, J.N. Reddy, A finite-element model for piezoelectric composite laminates, Smart Materials and Structures 6 (1997) 583–591.

[14] K.Y. Sze, Y.S. Pan, Hybrid finite element models for piezoelectric materials, Journal of Sound and Vibration 226 (1999) 519–547.

[15] C. Cao, Q.H. Qin, A. Yu, Hybrid fundamental-solution-based FEM for piezoelectric materials, Computational Mechanics 50 (2012) 397–412.

[16] Q.H. Qin, Variational formulations for TFEM of piezoelectricity, International Journal of Solids and Structures 40 (2003) 6335–6346.

[17] Q.H. Qin, Solving anti-plane problems of piezoelectric materials by the Trefftz finite element approach, Computational Mechanics 31 (2003) 461–468.

[18] J.S. Lee, L.Z. Jiang, A boundary integral formulation and 2D fundamental solution for piezoelastic media, Mechanics Research Communications 21 (1994) 47–54.

[19] H.J. Ding, G.Q. Wang, W.Q. Chen, A boundary integral formulation and 2D fundamental solutions for piezoelectric media, Computer Methods in Applied Mechanics and Engineering 158 (1998) 65–80.

[20] Q.H. Qin, Thermoelectroelastic analysis of cracks in piezoelectric half-plane by BEM, Computational Mechanics 23 (1999) 353–360.

[21] Q.H. Qin, Material properties of piezoelectric composites by BEM and homogenization method, Composite Structures 66 (2004) 295–299.

[22] G.S.A. Fam, Y.F. Rashed, An efficient meshless technique for the solution of transversely isotropic two-dimensional piezoelectricity, Computers and Mathematics with Applications 69 (2015) 438–454.

[23] W.A. Yao, H. Wang, Virtual boundary element integral method for 2-D piezoelectric media, Finite Elements in Analysis and Design 41 (2005) 875–891.

[24] W.A. Yao, H. Wang, Virtual boundary element-equivalent collocation method for plane piezoelectric materials (in Chinese), Chinese Journal of Computational Mechanics 22 (2005) 42–46.

[25] H.J. Ding, G. Wang, J. Liang, General and fundamental solutions of plane piezoelectroelastic problem, Acta Mechanica Sinica 28 (1996) 441–448.

[26] C.F. Gao, D.M. Cui, The fundamental solutions for the plane problem in piezoelectric medium, Chinese Journal of Applied Mechanics 16 (1999) 140–143.

[27] H.J. Ding, G.Q. Wang, W.Q. Chen, Fundamental solutions for plane problem of piezoelectric materials, Science in China - Series E: Technological Sciences 40 (1997) 331.

[28] H.J. Ding, J. Liang, The fundamental solutions for transversely isotropic piezoelectricity and boundary element method, Computers and Structures 71 (1999) 447–455.

[29] H. Sosa, Plane problems in piezoelectric media with defects, International Journal of Solids and Structures 28 (1991) 491–505.

[30] J.S. Lee, Boundary element method for electroelastic interaction in piezoceramics, Engineering Analysis with Boundary Elements 15 (1995) 321–328.

Chapter 8

Meshless analysis of heat transfer in heterogeneous media

Chapter outline

Methods of Fundamental Solutions in Solid Mechanics. https://doi.org/10.1016/B978-0-12-818283-3.00008-7

8.1 Introduction

The procedures presented in the previous chapters for the solutions of mechanical problems can be extended to a variety of physical problems, such as the heat conduction, seepage, or torsion of prismatic shafts, if more detailed attention is given to deal with a particular class of such situations, heat conduction, which is governed by the well-known Laplace or Poisson's equations. For example, the method of fundamental solutions (MFS) has been applied to steady-state heat conduction in isotropic or anisotropic materials [1–3] and transient heat conduction problems [4–6]. In the temperature approximation, the steady-state or transient fundamental solution is introduced to satisfy the governing partial differential equation. However, for the problems without explicit expressions of fundamental solution [7–10], the standard MFS cannot be applied directly, and other techniques have to be incorporated to extend the applicability of the MFS, including the analog equation method (AEM) and the radial basis function (RBF) interpolation, as introduced in the previous chapters.

The present chapter is concerned with applications of the extended MFS formulations for the solutions of the generalized partial differential equations with inhomogeneous terms that may appear in two types of heat conduction taken into consideration, one is the steady-state heat transfer in anisotropic material, and another is the transient heat conduction. Two-dimensional examples are provided to illustrate the formulations developed in this chapter, and their results are also presented.

8.2 Basics of heat transfer

8.2.1 Energy balance equation

Regarding the two-dimensional heat conduction problem referred to the Cartesian coordinates x and y given in the domain Ω, as shown in Fig. 8.1,

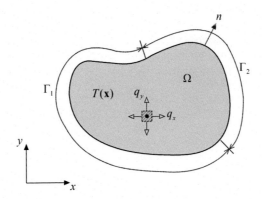

FIGURE 8.1 Schematics of two-dimensional heat conduction in the domain Ω.

the rate of heat energy per unit area [units: J/s/m²], named as the heat flux, at a field point $\mathbf{x} = (x, y)$ can be written in terms of its Cartesian components as

$$\mathbf{q} = [q_x \quad q_y]^{\mathrm{T}} \tag{8.1}$$

The governing equation of heat transfer can be derived by considering the energy balance of a microelement around a point $\mathbf{x}$, as shown in Fig. 8.2. The energy balance principle of the problem requires the following [11]:

Energy in + Energy generation = Energy out + Change of stored energy
$$\tag{8.2}$$

where the term "energy" refers to the heat power [units: J], which would be expressed as

$$\text{Energy} = (\text{heat flux}) \times (\text{area}) \times (\text{time}) \tag{8.3}$$

Therefore, if the thickness of the microelement is assumed to be 1, the energy in and out along the x-direction in the time interval dt can be written as

Energy in along x − direction: $\quad q_x \times \mathrm{d}y \times 1 \times \mathrm{d}t$

Energy out along x − direction: $\quad \left(q_x + \dfrac{\partial q_x}{\partial x} \mathrm{d}x \right) \times \mathrm{d}y \times 1 \times \mathrm{d}t$ $\tag{8.4}$

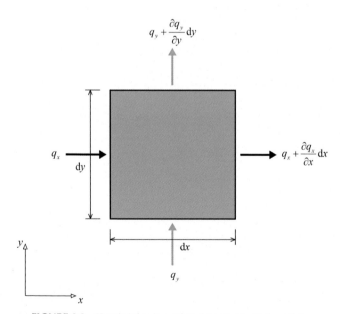

FIGURE 8.2   Heat balance in a microelement around the point $\mathbf{x}$.

Similarly, the energy in and out along the y-direction in the time interval dt can be given by

$$\text{Energy in along } y - \text{direction:} \quad q_y \times dx \times 1 \times dt$$

$$\text{Energy out along } y - \text{direction:} \quad \left(q_y + \frac{\partial q_y}{\partial y} dy\right) \times dx \times 1 \times dt \tag{8.5}$$

Besides, the generated energy caused by the interior heat source in the microelement can be written as

$$\text{Energy generation:} \quad Q \times dx \times dy \times 1 \times dt \tag{8.6}$$

where Q denotes the rate of heat generation per unit volume [units: J/s/m^3].

Moreover, a change of stored energy per unit mass is proportional to the change of temperature, so the change of stored energy in the microelement body can be given by the following:

$$\text{Change of stored energy:} \quad c \times (\rho \times dx \times dy \times 1) \times du \tag{8.7}$$

where c is the specific heat capacity of material [units: J/kg/K], ρ is the material density [units: kg/m^3], and du is the temperature change within the time dt [units: K].

Putting all terms in Eqs. (8.4)–(8.7) together produces

$$q_x dy dt + q_y dx dt + Q dx dy dt$$
$$= \left(q_x + \frac{\partial q_x}{\partial x} dx\right) dy dt + \left(q_y + \frac{\partial q_y}{\partial y} dy\right) dx dt + \rho c \, du \, dx dy \tag{8.8}$$

from which we have

$$\rho c \, du \, dx dy = -\frac{\partial q_x}{\partial x} dx dy dt - \frac{\partial q_y}{\partial y} dy dx dt + Q dx dy dt \tag{8.9}$$

Finally, dividing by the term $dx dy dt$ in Eq. (8.9) yields

$$\rho c \frac{du}{dt} = -\frac{\partial q_x}{\partial x} - \frac{\partial q_y}{\partial y} + Q \tag{8.10}$$

or

$$\rho c \frac{\partial u(\mathbf{x}, t)}{\partial t} = \frac{\partial q_x(\mathbf{x}, t)}{\partial x} - \frac{\partial q_y(\mathbf{x}, t)}{\partial y} + Q(\mathbf{x}, t), \quad \mathbf{x} \in \Omega \tag{8.11}$$

Specially, for the steady-state heat conduction, the heat balance equation can be obtained by removing the left term in Eq. (8.10) [11] as follows:

$$\frac{\partial q_x(\mathbf{x})}{\partial x} + \frac{\partial q_y(\mathbf{x})}{\partial y} = Q(\mathbf{x}), \quad \mathbf{x} \in \Omega \tag{8.12}$$

Introducing the gradient operator,

$$\nabla = \begin{bmatrix} \dfrac{\partial}{\partial x} & \dfrac{\partial}{\partial y} \end{bmatrix}^{\mathrm{T}} \tag{8.13}$$

we can rewrite Eq. (8.12) in matrix notation as

$$\nabla^{\mathrm{T}} \mathbf{q}(\mathbf{x}) = Q(\mathbf{x}), \quad \mathbf{x} \in \Omega \tag{8.14}$$

8.2.2 Fourier's law

Generally, the heat flux is related to the gradient of temperature variable u through the following Fourier's law of heat conduction:

$$\mathbf{q} = \begin{Bmatrix} q_x \\ q_y \end{Bmatrix} = - \begin{bmatrix} k_{xx} & k_{xy} \\ k_{xy} & k_{yy} \end{bmatrix} \begin{Bmatrix} \dfrac{\partial u}{\partial x} \\ \dfrac{\partial u}{\partial y} \end{Bmatrix} = -\mathbf{k}\nabla u \tag{8.15}$$

where $\mathbf{k}$ is a symmetric thermal conductivity matrix of anisotropic material,

$$\mathbf{k} = \begin{bmatrix} k_{xx} & k_{xy} \\ k_{xy} & k_{yy} \end{bmatrix} \tag{8.16}$$

which can be a constant matrix in homogeneous materials or a function of Cartesian coordinates in heterogeneous materials. In the generalized two-dimensional heat conduction, the most significant quantity to characterize the anisotropy of a medium is the determinant of the thermal conductivity matrix $\mathbf{k}$, that is,

$$\det(\mathbf{k}) = k_{xx}k_{yy} - k_{xy}^2 > 0 \tag{8.17}$$

The smaller the value of $\det(\mathbf{k})$ is, the more asymmetric the temperature field and the behavior of heat flux vectors are, and the more difficult the numerical calculation will be [1].

8.2.3 Governing equation

The final governing partial differential equation for the temperature variable can be obtained by substituting Eq. (8.15) into the heat balance Eq. (8.14) as

$$\nabla^{\mathrm{T}}(\mathbf{k}\nabla u) + Q = 0 \tag{8.18}$$

In our analysis, the temperature u is determined in a given domain Ω bounded by a boundary Γ.

Through the matrix expansion, the governing Eq. (8.18) can be rewritten as

$$\nabla^{\mathrm{T}}\mathbf{k}\nabla u + \mathbf{k}\nabla^{\mathrm{T}}\nabla u + Q = 0 \tag{8.19}$$

which can be reduced to

$$\mathbf{k}\nabla^{\mathrm{T}}\nabla u + Q = 0 \tag{8.20}$$

if the thermal conductivity matrix is independent of the spatial variable $\mathbf{x}$, i.e., it is for homogeneous materials.

Specially, for an isotropic material, the thermal conductivity matrix can be written as

$$\mathbf{k} = k \begin{bmatrix} 1 & 0 \\ 0 & 1 \end{bmatrix} = k\mathbf{I} \tag{8.21}$$

which leads to the simplest form of the governing partial differential Eq. (8.18), that is,

$$\nabla^T(k\mathbf{I}\nabla u) + Q = \nabla^T(k\nabla u) + Q = \nabla^T k\nabla u + k\nabla^T\nabla u + Q = 0 \tag{8.22}$$

Furthermore, if the material is homogeneous, that is, $k = \text{const}$, Eq. (8.22) can be simplified as

$$k\nabla^2 u + Q = 0 \tag{8.23}$$

8.2.4 Boundary conditions

On the boundary of a solution domain we usually encounter one of the following conditions.

Dirichlet condition:

$$u(\mathbf{x}) = \bar{u}(\mathbf{x}), \quad \mathbf{x} \in \Gamma_1 \tag{8.24}$$

Neumann condition:

$$q_n(\mathbf{x}) = \bar{q}(\mathbf{x}), \quad \mathbf{x} \in \Gamma_2 \tag{8.25}$$

where $\bar{T}$ and $\bar{q}$ are specified values on the boundary Γ_1 and Γ_1, respectively, and $\Gamma = \Gamma_1 + \Gamma_2$. In Eq. (8.25), q_n denotes the normal heat flux defined as

$$q_n = \mathbf{n}^T\mathbf{q} = -\mathbf{n}^T k\nabla u \tag{8.26}$$

where $\mathbf{n} = \begin{bmatrix} n_x & n_y \end{bmatrix}^T$ is a unit normal vector of direction cosines to the boundary.

8.2.5 Thermal conductivity matrix

The general definition of the thermal conductivity matrix $\mathbf{k}$ in Eq. (8.16) is made for an anisotropic material with respect to a given set of global coordinate axes x and y. In engineering, when a material is orthotropic, it is always possible to determine its global material thermal conductivity matrix through a diagonal matrix $\tilde{\mathbf{k}}$

$$\tilde{\mathbf{k}} = \begin{bmatrix} k_{\tilde{x}\tilde{x}} & 0 \\ 0 & k_{\tilde{y}\tilde{y}} \end{bmatrix} \tag{8.27}$$

which is defined with respect to the a local coordinate system ($\tilde{x}$ and $\tilde{y}$ as shown in Fig. 8.3). The local coordinate axes are usually selected the same as those of material principle axes.

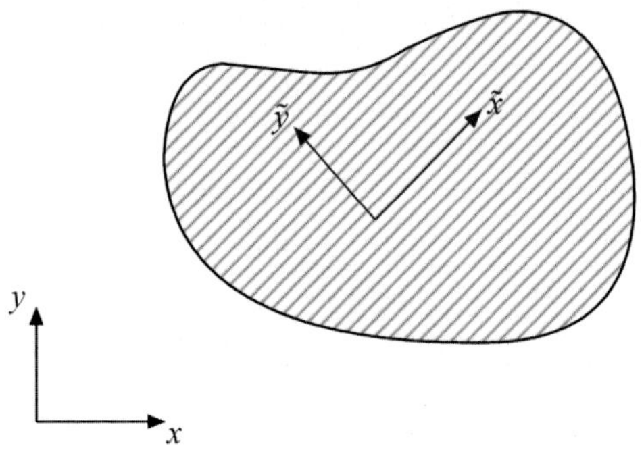

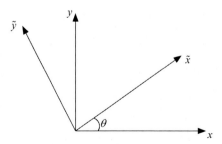

FIGURE 8.3   Local material principal axes.

The orientation of the local axes $\widetilde{x}$ and $\widetilde{y}$ can be determined by a transformation matrix T with respect to the global coordinates (see Appendix B for details):

$$\widetilde{\mathbf{x}} = \left\{ \begin{matrix} \widetilde{x} \\ \widetilde{y} \end{matrix} \right\} = \mathbf{T} \left\{ \begin{matrix} x \\ y \end{matrix} \right\} = \mathbf{Tx} \tag{8.28}$$

in which **T** is the transformation matrix defined as

$$\mathbf{T} = \begin{bmatrix} \cos(\widetilde{x}, x) & \cos(\widetilde{x}, y) \\ \cos(\widetilde{y}, x) & \cos(\widetilde{y}, y) \end{bmatrix} \tag{8.29}$$

where $\cos(\widetilde{x}, x)$ is the cosine of the angle between the $\widetilde{x}$ direction and the x-direction, and other cosines can be similarly defined.

If the angle between the $\widetilde{x}$ direction and the x-direction is denoted as θ, as indicated in Fig. 8.4, the transformation matrix can be rewritten as

$$\mathbf{T} = \begin{bmatrix} \cos\theta & \sin\theta \\ -\sin\theta & \cos\theta \end{bmatrix} \tag{8.30}$$

from which we have

$$\mathbf{T}^{\mathrm{T}}\mathbf{T} = \mathbf{T}\mathbf{T}^{\mathrm{T}} = \mathbf{I} \tag{8.31}$$

FIGURE 8.4   Coordinate transformation.

or

$$\mathbf{T}^{-1} = \mathbf{T}^{\mathrm{T}} \tag{8.32}$$

where $\mathbf{I}$ is a 2×2 identity matrix.

Correspondingly, the governing partial differential equation for these local principal axes can be written as

$$-\widetilde{\nabla}^{\mathrm{T}}\left(\widetilde{\mathbf{k}}\widetilde{\nabla}T\right) + Q = 0 \tag{8.33}$$

where

$$\widetilde{\nabla} = \left[\frac{\partial}{\partial\widetilde{x}} \ \frac{\partial}{\partial\widetilde{y}}\right]^{\mathrm{T}} \tag{8.34}$$

Using the coordinate transformation and the chain rule, we have

$$\widetilde{\nabla} = \mathbf{T}\nabla \tag{8.35}$$

or alternatively

$$\nabla = \mathbf{T}^{\mathrm{T}}\widetilde{\nabla} \tag{8.36}$$

Substituting Eq. (8.35) into the governing partial differential equation with respect to the local axes gives

$$\begin{aligned}
-(\mathbf{T}\nabla)^{\mathrm{T}}\left(\widetilde{\mathbf{k}}\mathbf{T}\nabla T\right) + Q \\
= -\nabla^{\mathrm{T}}\mathbf{T}^{\mathrm{T}}\left(\widetilde{\mathbf{k}}\mathbf{T}\nabla T\right) + Q \\
= -\nabla^{\mathrm{T}}\left(\mathbf{T}^{\mathrm{T}}\widetilde{\mathbf{k}}\mathbf{T}\nabla T\right) + Q \\
= 0
\end{aligned} \tag{8.37}$$

Comparing Eq. (8.37) with Eq. (8.18) gives

$$\mathbf{k} = \mathbf{T}^{\mathrm{T}}\widetilde{\mathbf{k}}\mathbf{T} \tag{8.38}$$

or alternatively

$$\widetilde{\mathbf{k}} = \mathbf{T}\mathbf{k}\mathbf{T}^{\mathrm{T}} \tag{8.39}$$

8.3 Solution procedure of general steady-state heat transfer

8.3.1 Solution procedure

To make the formulations for the general steady-state heat transfer in heterogeneous materials more compact, the following linear differential operator is introduced:

$$\Re(u) = \nabla^{\mathrm{T}}\mathbf{k}\nabla u + \mathbf{k}\nabla^{\mathrm{T}}\nabla u \tag{8.40}$$

so the governing Eq. (8.19) can be rewritten as

$$\Re[u(\mathbf{x})] + Q(\mathbf{x}) = 0 \tag{8.41}$$

8.3.1.1 Analog equation method

The boundary value problems (BVPs) defined by the governing Eq. (8.41) and the boundary conditions (8.24) and (8.25) can be converted into a standard Poisson type equation using the analog equation method [12]. To do this $u(\mathbf{x})$ is assumed to be the sought solution to the BVPs, which is a continuously differentiable function with up to two orders in the solution domain Ω. If the Laplacian operator is applied to this function, we have

$$\nabla^2 u(\mathbf{x}) = b(\mathbf{x}) \tag{8.42}$$

Eq. (8.42) indicates that the solution of Eq. (8.41) can be obtained by solving the linear Eq. (8.42) subjected to the same boundary conditions (8.24) and (8.25), if the fictitious source distribution $b(\mathbf{x})$ is known. The solution procedure is detailed next.

Since the fictitious Laplace Eq. (8.42) is linear, its solution can be written as a summation of the homogeneous solution $u^h(\mathbf{x})$ and the inhomogeneous solution or particular solution $u^p(\mathbf{x})$, that is,

$$u(\mathbf{x}) = u^h(\mathbf{x}) + u^p(\mathbf{x}) \tag{8.43}$$

Accordingly, the homogeneous solution $u^h(\mathbf{x})$ and the particular solution $u^p(\mathbf{x})$ should, respectively, satisfy

$$\nabla^2 u^h(\mathbf{x}) = 0 \tag{8.44}$$

and

$$\nabla^2 u^p(\mathbf{x}) = b(\mathbf{x}) \tag{8.45}$$

whose expressions can be respectively derived by the MFS and the RBF interpolation, as implemented next.

8.3.1.2 Particular solution

The next step of the proposed meshless approach is to determine the particular solution by the RBF approximation. To this end, the right-hand term of Eq. (8.45) can be approximated by

$$b(\mathbf{x}) = \sum_{m=1}^{M} \alpha_m f_m(\mathbf{x}) \tag{8.46}$$

where M is the total number of interpolation points inside the domain and on the boundary, α_m are coefficients to be determined, and $f_m(\mathbf{x})$ is a set of approximating basis functions. Such an approximation can always be written in the matrix form

$$b(\mathbf{x}) = \mathbf{f}^T \boldsymbol{\alpha} \tag{8.47}$$

where

$$\mathbf{f}^{\mathrm{T}} = [f_1(\mathbf{x}) \quad f_2(\mathbf{x}) \quad \cdots \quad f_M(\mathbf{x})] \tag{8.48}$$

$$\boldsymbol{\alpha}^{\mathrm{T}} = [\alpha_1 \quad \alpha_2 \quad \cdots \quad \alpha_M] \tag{8.49}$$

Similarly, the particular solution $u^P(\mathbf{x})$ is approximated by the following form:

$$u^P(\mathbf{x}) = \sum_{m=1}^{M} \alpha_m \widehat{u}_m(\mathbf{x}) = \widehat{\mathbf{u}}^{\mathrm{T}} \boldsymbol{\alpha} \tag{8.50}$$

where

$$\widehat{\mathbf{u}}^{\mathrm{T}}(\mathbf{x}) = \left[\widehat{u}_1(\mathbf{x}) \quad \widehat{u}_2(\mathbf{x}) \quad \cdots \quad \widehat{u}_M(\mathbf{x}) \right] \tag{8.51}$$

is the vector consisting of a set of particular solutions $\widehat{u}_m(\mathbf{x})$. Because the particular solution $u^P(\mathbf{x})$ is required to satisfy Eq. (8.45), the precondition for such approximation is that the following relation exists:

$$\nabla^2 \widehat{u}_m(\mathbf{x}) = f_m(\mathbf{x}) \tag{8.52}$$

The effectiveness and accuracy of the interpolation depends on the choice of the approximating basis functions $f_m(\mathbf{x})$. Here, the power-type RBF in terms of Euclidean distance r_j are chosen to be approximating basis functions [13,14].

$$f_j(\mathbf{x}) = 1 + r_j^n(\mathbf{x}), \quad n = 1, 3, 5, \ldots \tag{8.53}$$

where $r_j(\mathbf{x}) = r(\mathbf{x}, \mathbf{x}_j) = \|\mathbf{x} - \mathbf{x}_j\|$ denotes the Euclidean distance from the source point $\mathbf{x}_j$ to the field point $\mathbf{x}$. In Eq. (8.53), only odd powers are allowed, because even powers of r_j are not RBFs, and with even powers artificially created singularities may be encountered in some cases [15].

Substituting Eq. (8.53) into Eq. (8.52), we have the following [3]:

$$\widehat{u}_j(\mathbf{x}) = \frac{r_j^2(\mathbf{x})}{4} + \frac{r_j^5(\mathbf{x})}{25} \tag{8.54}$$

for $n = 3$, and

$$\widehat{u}_j(\mathbf{x}) = \frac{r_j^2(\mathbf{x})}{4} + \frac{r_j^3(\mathbf{x})}{9} \tag{8.55}$$

for $n = 1$.

Since the inhomogeneous term $b(\mathbf{x})$ is an unknown function depending on the unknown function $u(\mathbf{x})$, the coefficients α_j cannot be determined directly through solving Eq. (8.47) at $\mathbf{x}_m$ ($m = 1, 2, \ldots, M$). This obstacle can be tackled in an indirect way, as described later.

8.3.1.3 Homogeneous solution

To obtain an approximated homogeneous solution of Laplace Eq. (8.44), N virtual point sources $\mathbf{y}_j (j = 1, 2, ..., N)$ are placed outside the solution domain. The homogeneous solution at a point of interest $\mathbf{x} \in \Omega$ can then be approximated by a linear combination of the fundamental solutions of Laplace equation:

$$u^h(\mathbf{x}) = \sum_{i=1}^{N} u^*(\mathbf{x}, \mathbf{y}_i) \varphi_i, \quad \mathbf{x} \in \Omega, \ \mathbf{y}_i \notin \Omega \tag{8.56}$$

where

$$u^*(\mathbf{x}, \mathbf{y}_i) = \frac{1}{2\pi} \ln \frac{1}{r(\mathbf{x}, \mathbf{y}_i)} \tag{8.57}$$

In matrix notation, Eq. (8.56) can be rewritten as

$$u^h(\mathbf{x}) = \mathbf{U}^{*T} \boldsymbol{\varphi} \tag{8.58}$$

where

$$\mathbf{U}^* = [u^*(\mathbf{x}, \mathbf{y}_1) \quad u^*(\mathbf{x}, \mathbf{y}_2) \quad ... \quad u^*(\mathbf{x}, \mathbf{y}_N)]^T$$
$$\boldsymbol{\varphi} = [\varphi_1 \quad \varphi_2 \quad ... \quad \varphi_M]^T \tag{8.59}$$

8.3.1.4 Approximated full solution

According to the analysis before, the full solution $u(\mathbf{x})$ we are seeking can be given by

$$u(\mathbf{x}) = u^h(\mathbf{x}) + u^p(\mathbf{x}) = \mathbf{U}^T(\mathbf{x}) \boldsymbol{\varphi} + \boldsymbol{\Phi}^T(\mathbf{x}) \boldsymbol{\alpha} = [\mathbf{U}^T(\mathbf{x}) \quad \boldsymbol{\Phi}^T(\mathbf{x})] \begin{bmatrix} \boldsymbol{\varphi} \\ \boldsymbol{\alpha} \end{bmatrix} = \mathbf{U}(\mathbf{x}) \boldsymbol{\beta} \tag{8.60}$$

where

$$\mathbf{U}(\mathbf{x}) = [\mathbf{U}^T(\mathbf{x}) \quad \boldsymbol{\Phi}^T(\mathbf{x})] \tag{8.61}$$

$$\boldsymbol{\beta} = \begin{bmatrix} \boldsymbol{\varphi} \\ \boldsymbol{\alpha} \end{bmatrix} \tag{8.62}$$

Then, the heat flux component can be obtained as

$$\mathbf{q}(\mathbf{x}) = -\mathbf{k}\nabla u(\mathbf{x}) = -\mathbf{k}\nabla\mathbf{U}(\mathbf{x})\boldsymbol{\beta} \tag{8.63}$$

Further, the boundary normal heat flux can be written as

$$q_n = -\mathbf{n}^T \mathbf{k}\nabla u = -\mathbf{n}^T \mathbf{k}\nabla\mathbf{U}\boldsymbol{\beta} \tag{8.64}$$

8.3.1.5 Construction of solving equations

To determine the unknowns α_j and φ_i in the full temperature solution (8.60), the same N boundary collocations are chosen. The satisfaction of u defined in Eq. (8.60) to the real governing Eq. (8.41) at M interpolation points in the domain Ω yields

$$\Re[\mathbf{U}(\mathbf{x}_i)\boldsymbol{\beta}] + Q(\mathbf{x}_i) = \Re[\mathbf{U}(\mathbf{x}_i)]\boldsymbol{\beta} + Q(\mathbf{x}_i) = 0, \quad \mathbf{x}_i \in \Omega, \ i = 1, 2, \ldots, M \tag{8.65}$$

Besides, the satisfaction of the specific boundary conditions (8.24) and (8.25) at N boundary collocations for the temperature solution (8.60) and the normal heat flux solution (8.64) means

$$\begin{aligned} \mathbf{U}(\mathbf{x}_j)\boldsymbol{\beta} &= \bar{u}(\mathbf{x}_j) \\ [-\mathbf{n}^T\mathbf{k}\nabla\mathbf{U}(\mathbf{x}_j)]\boldsymbol{\beta} &= \bar{q}_n(\mathbf{x}_j) \end{aligned} \tag{8.66}$$

Combining Eqs. (8.65) and (8.66), we obtain a linear system of equations

$$\begin{bmatrix} \Re[\mathbf{U}(\mathbf{x}_i)] \\ \mathbf{U}(\mathbf{x}_j) \\ -\mathbf{n}^T\mathbf{k}\nabla\mathbf{U}(\mathbf{x}_j) \end{bmatrix} \boldsymbol{\beta} = \begin{bmatrix} -Q(\mathbf{x}_i) \\ \bar{u}(\mathbf{x}_j) \\ \bar{q}_n(\mathbf{x}_j) \end{bmatrix} \tag{8.67}$$

or that in abbreviated form

$$\mathbf{A}_{(M+N)\times(M+N)}\boldsymbol{\beta} = \mathbf{f} \tag{8.68}$$

from which the unknown coefficient vector $\boldsymbol{\beta}$ can be determined. Once these unknown coefficients are determined, the temperature solution u and the heat flux component at any field point $\mathbf{x} \in \Omega$ can be evaluated using Eqs. (8.60) and (8.63).

8.3.2 Results and discussion

To demonstrate the efficiency and accuracy of the proposed method, several benchmark numerical examples [16−18] are considered and their results are compared with the available analytical results.

8.3.2.1 Isotropic heterogeneous square plate

As the first example, let us consider a 100×100 m isotropic square region illustrated in Fig. 8.5. The top and bottom boundaries are insulated. The left wall is assigned with the temperature of 200°C and the right wall is assigned with the temperature of 100°C.

The spatial variation of the thermal conductivity is taken in cubic order in x-direction, as

$$k(x, y) = \left(1 + \frac{x}{100}\right)^3 \tag{8.69}$$

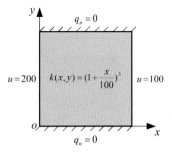

FIGURE 8.5   Geometry of square plate and boundary conditions.

The problem is two-dimensional, and an its analytical solution is

$$u(x, y) = \frac{800}{6} \left[\frac{1}{\left(1 + \dfrac{x}{100}\right)^2} + \frac{1}{2} \right] \tag{8.70}$$

To compare the different situations more easily, we introduce a measurement to judge the overall performance of each situation. To this end, we define the maximum percent relative error between two functions f and g, which represent the analytical and numerical solutions, respectively, as

$$\text{MRE} = \max_{(x,y) \in \Omega} \left| \frac{f - g}{f} \right| \times 100\% \tag{8.71}$$

where Ω is the computational domain within which both functions are defined. It is evident that the smaller the maximum relative error is, the better an approach is.

For this example, Figs. 8.6 and 8.7 depict the profiles of the RBF interpolation points in the square and the virtual source points outside the square, respectively. Different situations are analyzed to demonstrate the accuracy and efficiency of the proposed method.

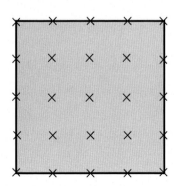

FIGURE 8.6   Profile of the RBF interpolation points in the square.

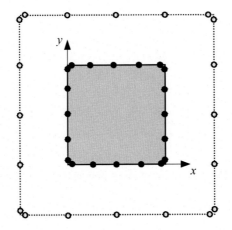

FIGURE 8.7 Profiles of virtual source points and boundary nodes on the virtual and real boundaries.

Firstly, the influence of the location of fictitious boundary is investigated. Fig. 8.8 depicts the variation of the maximum percent relative error for temperature within the square domain when the similarity ratio between the virtual and the real boundary defined in Eq. (1.50) varies from 1.8 to 4, with $N = 20$ and $M = 25$. Due to the singularity of the fundamental solutions, accuracy will decrease when the similarity ratio approaches 1, as can be seen from Fig. 8.8. We can also see that the similarity ratio can be varied within a certain range to obtain stable computing accuracy. Generally, the similarity ratio can be selected in the range of 1.8−4.0.

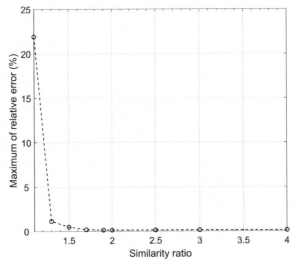

FIGURE 8.8 Variation of MRE for temperature when the similarity ratio varies from 1.8 to 4, with $N = 20$ and $M = 25$.

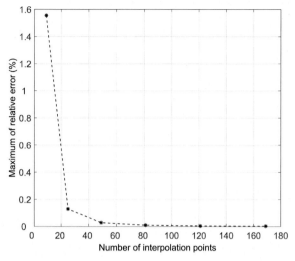

FIGURE 8.9   Variation of MRE for temperature when the number of the RBF interpolation points *M* changes.

Fig. 8.9 shows the variation of maximum of percent relative error for temperature within the square domain when the number of RBF interpolation points varies from 20 to 160 and the similarity ratio is 3.0, from which we can see that the greater the number of interpolation points is, the smaller the maximum relative error is.

Fig. 8.10 gives the variation of the maximum of percent relative error for temperature when the number of fictitious source points outside the domain

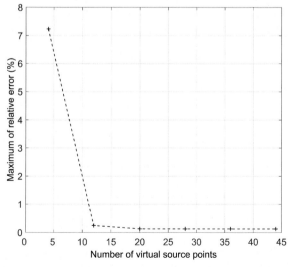

FIGURE 8.10   Variation of MRE for temperature when the number of virtual source points varies from 5 to 45 with *M* = 25.

varies from 5 to 45 with $M = 25$. In this case, no obvious increase in accuracy is observed when the number of virtual source points exceeds 20. The reasons for this effect may lie in the simpler boundary shape and the smaller number of the RBF interpolation points.

8.3.2.2 Isotropic heterogeneous circular disc

Here, we consider an isotropic disc whose radius is $R = 1$ m. The thermal conductivity is assumed to vary biquadratically as

$$k(x, y) = (2x + y + 2)^2 \tag{8.72}$$

Correspondingly, an analytical expression for the temperature field satisfying the governing heat conduction equation can be given as

$$u(x, y) = \frac{6x^2 - 6y^2 + 20xy + 30}{2x + y + 2} \tag{8.73}$$

which is also used to impose the Dirichlet boundary conditions shown in Fig. 8.11.

Additionally, the analytical temperature solution gives the follow heat flux solutions in the circular domain

$$q_x = -12x^2 - 12xy - 24x - 32y^2 - 40y + 60$$
$$q_y = 24xy + 6y^2 + 24y - 34x^2 - 40x + 30 \tag{8.74}$$

This nonhomogeneous problem is solved using 36 fictitious source points outside of the domain (shown in Fig. 8.12) and 181 interpolation points (shown in Fig. 8.13). The distribution of temperature on the circular line $r = 0.5$ inside the disc is shown in Fig. 8.14, while the boundary heat fluxes are depicted in Fig. 8.15. It is clear from these figures that good agreement between the meshless results and the analytical solutions is achieved with relatively fewer collocation points.

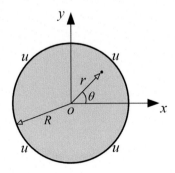

FIGURE 8.11 Geometry of the disc domain and boundary conditions.

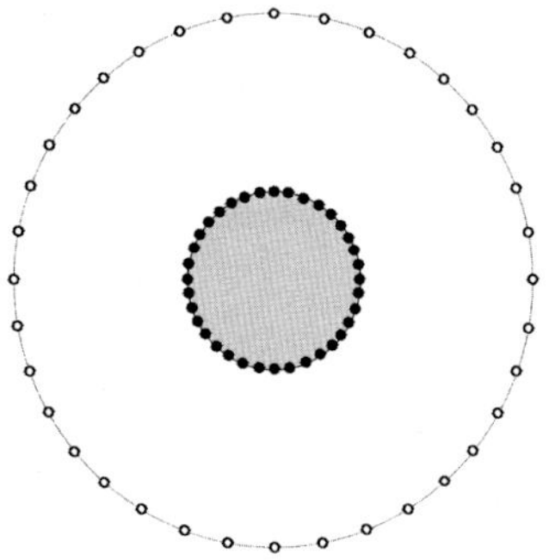

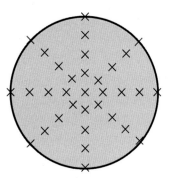

FIGURE 8.12   Illustration of collocation points on the virtual and real boundaries.

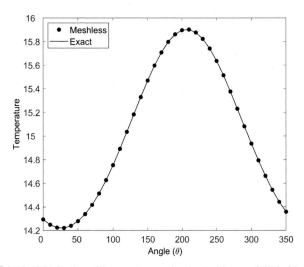

FIGURE 8.13   Illustration of interpolation points in the circular domain.

FIGURE 8.14   Distribution of temperature on the circular line $r = 0.05$ inside the disc.

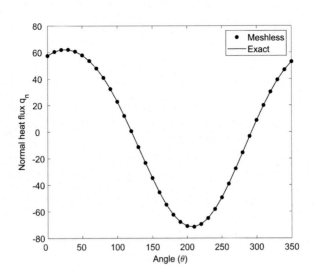

FIGURE 8.15 Distribution of normal heat flux on the boundary of the disc.

8.3.2.3 Anisotropic homogeneous circular disc

To illustrate numerical results for a typical anisotropic problem, we consider a 2D anisotropic medium with the thermal conductivity tensor given by

$$k_{11} = 5.0, \ k_{22} = 1.0, \text{ and } k_{12} = k_{21} = 2.0.$$

The Dirichlet problems are solved in the plane circular domain $\Omega = \{(x, y): x^2 + y^2 \leq 1\}$, i.e., the 2D disc of radius unity. The analytical temperature distribution to be retrieved is given by

$$u(x, y) = \frac{x^3}{5} - x^2 y + xy^2 + \frac{y^3}{5} \tag{8.75}$$

which is also used to impose the Dirichlet boundary conditions on the circular boundary.

The corresponding internal heat source and heat flux components can be given by

$$Q(x, y) = \frac{4}{5} y \tag{8.76}$$

$$\begin{aligned} q_x &= -x^2 + 6xy - 6.2y^2 \\ q_y &= -0.2x^2 + 2xy - 2.6y^2 \end{aligned} \tag{8.77}$$

In computation, the similar ratio is selected to be equal to 3.0. The number of fictitious source points on the virtual boundary is 36, and there are 33 interpolation points to be used for RBF approximation, including 25 internal points and eight boundary points. Fig. 8.16 shows the distribution of temperature at the points on the circles whose radii are equal to 0.5 and 1.0, respectively. We can see that a good agreement is reached between the

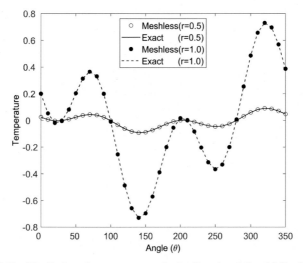

FIGURE 8.16   Distribution of temperature on circular lines ($r = 0.5$ and 1.0) within a disc.

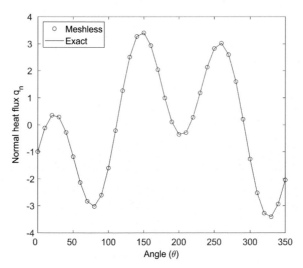

FIGURE 8.17   Distribution of normal heat flux on the boundary of the disc.

numerical results and the analytical solutions. Besides, Fig. 8.17 depicts the distribution of heat flux on the boundary, and we can see that the numerical results also agree well with the analytical solutions again.

8.3.2.4 Anisotropic heterogeneous hollow ellipse

As the final example in this chapter, let us consider an orthotropic material in a hollow ellipse (see Fig. 8.18) whose center is at $(0, 0)$, with its major axis of

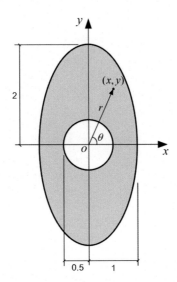

FIGURE 8.18 Anisotropic hollow ellipse and imposed boundary conditions.

2 m long in the y-direction and minor axis of 1 m long in the x-direction. The ellipse then encloses a circle of radius 0.5 m whose center is also at (0, 0) as well. The thermal conductivity is

$$k(x, y) = \begin{bmatrix} 2x + y + 5 & 0 \\ 0 & 3x + y + 7 \end{bmatrix} \tag{8.78}$$

The following temperature distribution

$$u(x, y) = 4x^2 + 10xy - 7y^2 + 20x + 18y \tag{8.79}$$

can be shown to satisfy the governing heat conduction equation with the conductivity given before. As such, it is used to impose the temperature boundary condition along the outer elliptical boundary and the normal heat flux boundary condition along the inner circular boundary.

Correspondingly, the analytical heat flux solution can be obtained by

$$\begin{aligned} q_x &= -(2x + y + 5)(8x + 10y + 20) \\ q_y &= -(3x + y + 7)(10x - 14y + 18) \end{aligned} \tag{8.80}$$

In the computation, 24 fictitious source points on the virtual boundary and the same number of collocation points on the real boundary are employed, as indicated in Fig. 8.19. The similar ratio is equal to 3.0 and 0.6, respectively. Additionally, 108 interior points and 72 boundary points are selected as interpolation points to perform RBF approximation (see Fig. 8.20).

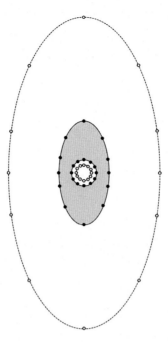

FIGURE 8.19   Illustration of collocation points on the virtual and real boundaries.

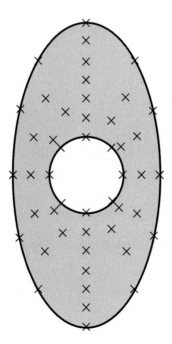

FIGURE 8.20   Illustration of interpolation points on the boundary and within the domain.

The numerical results of temperature and heat flux on the inner circular boundary and outer elliptic boundary are shown in Figs. 8.21−8.24 and they are compared with the analytical results, from which it can be seen that the proposed meshless method has good accuracy.

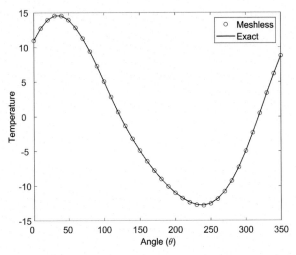

FIGURE 8.21 Comparison of temperature between numerical and exact results on the inner circular boundary.

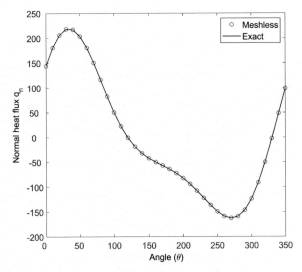

FIGURE 8.22 Comparison of normal heat flux between numerical and exact results on the inner circular boundary.

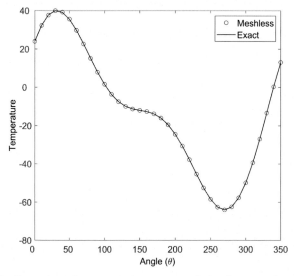

FIGURE 8.23 Comparison of temperature between numerical and exact results on the external elliptic boundary.

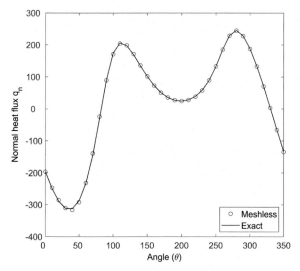

FIGURE 8.24 Comparison of normal heat flux between numerical and exact results on the external elliptic boundary.

In addition to the results at points on the inner and outer boundaries, the distribution of temperature at some internal points located on the ellipse, which has a semimajor axis of 1.25 m long and a semiminor axis of 0.75 m long, is also computed and presented in Fig. 8.25. It can be seen again from Fig. 8.25 that the proposed method has relatively high accuracy.

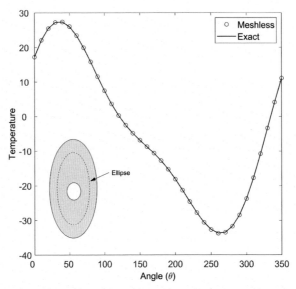

FIGURE 8.25 Comparison of temperature between numerical and exact results on the ellipse that has a semimajor axis of length 1.25 m and a semiminor axis of length 0.75 m.

8.4 Solution procedure of transient heat transfer

The preceding formulations have assumed that the temperature solution to the problem is independent of time. However, many problems like the melting of metal require the time-dependent solution, for both the boundary conditions and the partial differential equation. For such cases, the temperature solution requires satisfying the partial differential equation given by

$$\rho c \frac{\partial u(\mathbf{x}, t)}{\partial t} = \nabla^{\mathrm{T}}[\mathbf{k}\nabla u(\mathbf{x}, t)] + Q(\mathbf{x}, t) \tag{8.81}$$

where ρ is the density of material and c the specific heat per unit mass. Therefore, the product of them stands for the specific heat per unit volume. Besides, the time variable t is assumed to change in the interval $[0, T]$.

Correspondingly, the transient boundary conditions can be written in the form

$$u(\mathbf{x}, t) = \bar{u}(\mathbf{x}, t), \quad \mathbf{x} \in \Gamma_1 \tag{8.82}$$

$$q_n(\mathbf{x}, t) = \bar{q}(\mathbf{x}, t), \quad \mathbf{x} \in \Gamma_2 \tag{8.83}$$

In addition to the boundary conditions, it is necessary to provide the distribution of temperature at the initial time, i.e.,

$$u(\mathbf{x}, 0) = u_0(\mathbf{x}), \quad \mathbf{x} \in \Omega \tag{8.84}$$

where T_0 denotes the given initial temperature in the domain.

8.4.1 Solution procedure

8.4.1.1 Time marching scheme

For the transient heat transfer problems, to determine the variation of temperature field in the computational domain at any time instance, it is necessary to divide the time domain into finite numbers of time intervals [6]:

$$[0, T] = \{0, t_1, t_2, ..., t_n, t_{n+1}, ..., T\}$$

There are many ways to deal with the terms related to the time variable t in the governing Eq. (8.81) and the boundary conditions (8.82)−(8.83). Here, the so-called θ-method is employed, which can lead to the most commonly used algorithm for time treatment. In the θ-method, the first-order time derivative is first replaced by a simple difference as

$$\frac{\partial u(\mathbf{x}, t)}{\partial t} = \frac{u^{n+1}(\mathbf{x}) - u^n(\mathbf{x})}{\Delta t}, \quad t_n \leq t \leq t_{n+1}, \quad n = 0, 1, ... \tag{8.85}$$

where $u^n(\mathbf{x}) = u(\mathbf{x}, t^n)$ and $u^{n+1}(\mathbf{x}) = u(\mathbf{x}, t^{n+1})$ denote the temperatures at the time instance t^n and t^{n+1}, respectively, and $\Delta t = t^{n+1} - t^n$ is the specific time increment.

It is assumed that the temperature field $u^n(\mathbf{x}) = u(\mathbf{x}, t_n)$ is already known and then is used as an initial condition to evaluate the unknown solution at the next time instance t_{n+1}. In the θ-method, the temperature solution at the time $t \in [t_n, t_{n+1}]$ is written as

$$u(\mathbf{x}, t) = \theta u^n(\mathbf{x}) + (1 - \theta)u^{n+1}(\mathbf{x}), \quad t_n \leq t \leq t_{n+1} \tag{8.86}$$

where θ denotes the relaxation parameter, which is usually specified in the range $0 \leq \theta \leq 1$, and can be used to control the accuracy and stability of the algorithm. The most commonly used values of θ are 0, 1/2, and 1, which correspond to the forward explicit Euler method, the central implicit method, and the backward implicit Euler method. Among them, the backward Euler method is explicit and unconditionally stable.

In this study, the backward Euler method ($\theta = 1$) is used. Thus, the governing Eq. (8.81) becomes

$$\rho c \frac{u^{n+1}(\mathbf{x}) - u^n(\mathbf{x})}{\Delta t} = \nabla^T \left[\mathbf{k} \nabla u^{n+1}(\mathbf{x}) \right] + Q^n(\mathbf{x}) \tag{8.87}$$

from which we have

$$u^{n+1}(\mathbf{x}) - \nabla^T \left[\mathbf{k} \nabla u^{n+1}(\mathbf{x}) \right] \frac{\Delta t}{\rho c} = Q^n(\mathbf{x}) \frac{\Delta t}{\rho c} + u^n(\mathbf{x}) \tag{8.88}$$

or

$$u^{n+1}(\mathbf{x}) - \left[\nabla^T \mathbf{k} \nabla u^{n+1}(\mathbf{x}) + \mathbf{k} \nabla^T \nabla u^{n+1}(\mathbf{x}) \right] \frac{\Delta t}{\rho c} = Q^n(\mathbf{x}) \frac{\Delta t}{\rho c} + u^n(\mathbf{x}) \tag{8.89}$$

If we introduce the following differential operator

$$\widetilde{\Re}\left[u^{n+1}(\mathbf{x})\right] = u^{n+1}(\mathbf{x}) - \nabla^{\mathrm{T}}\left[k(\mathbf{x})\nabla u^{n+1}(\mathbf{x})\right]\frac{\Delta t}{\rho c}$$

$$= u^{n+1}(\mathbf{x}) - \left[\nabla^{\mathrm{T}}k(\mathbf{x})\nabla u^{n+1}(\mathbf{x}) + k(\mathbf{x})\nabla^{\mathrm{T}}\nabla u^{n+1}(\mathbf{x})\right]\frac{\Delta t}{\rho c} \tag{8.90}$$

then Eq. (8.89) can be rewritten as

$$\widetilde{\Re}\left[u^{n+1}(\mathbf{x})\right] = Q^n(\mathbf{x})\frac{\Delta t}{\rho c} + u^n(\mathbf{x}) \tag{8.91}$$

Besides, the boundary conditions at the time instance t^{n+1} can be rewritten from Eqs. (8.82) and (8.83) as

$$u^{n+1}(\mathbf{x}) = \bar{u}(\mathbf{x}, t^{n+1}), \quad \mathbf{x} \in \Gamma_1 \tag{8.92}$$

$$q_n^{(n+1)}(\mathbf{x}) = \bar{q}(\mathbf{x}, t^{n+1}), \quad \mathbf{x} \in \Gamma_2 \tag{8.93}$$

8.4.1.2 Approximated full solution

Following the procedure including the analog equation method, the method of fundamental solution, and the radial basis function interpolation for the steady-state heat conduction, the full solution at the time instance t^{n+1} of Eq. (8.91) can be approximated by

$$\begin{aligned} u^{n+1}(\mathbf{x}) &= u^{h,n+1}(\mathbf{x}) + u^{p,n+1}(\mathbf{x}) \\ &= \mathbf{U}^{\mathrm{T}}(\mathbf{x})\boldsymbol{\varphi}^{n+1} + \boldsymbol{\Phi}^{\mathrm{T}}(\mathbf{x})\boldsymbol{\alpha}^{n+1} \\ &= \left[\mathbf{U}^{\mathrm{T}}(\mathbf{x}) \quad \boldsymbol{\Phi}^{\mathrm{T}}(\mathbf{x})\right]\begin{bmatrix}\boldsymbol{\varphi}^{n+1} \\ \boldsymbol{\alpha}^{n+1}\end{bmatrix} \\ &= \mathbf{U}(\mathbf{x})\boldsymbol{\beta}^{n+1} \end{aligned} \tag{8.94}$$

where $\mathbf{U}^{\mathrm{T}}(\mathbf{x})$ and $\boldsymbol{\Phi}^{\mathrm{T}}(\mathbf{x})$ are the fundamental kernel vector and the particular solution kernel vector, which are same as those defined in Section 8.3.1, $\boldsymbol{\varphi}^{n+1}$ and $\boldsymbol{\alpha}^{n+1}$ are unknown vectors at the time instance t^{n+1} that appeared in the method of fundamental solution and the RBF interpolation, and

$$\mathbf{U}(\mathbf{x}) = \left[\mathbf{U}^{\mathrm{T}}(\mathbf{x}) \quad \boldsymbol{\Phi}^{\mathrm{T}}(\mathbf{x})\right] \tag{8.95}$$

$$\boldsymbol{\beta}^{n+1} = \begin{bmatrix}\boldsymbol{\varphi}^{n+1} \\ \boldsymbol{\alpha}^{n+1}\end{bmatrix} \tag{8.96}$$

Subsequently, the heat flux component at the time instance t^{n+1} can be derived by

$$\mathbf{q}(\mathbf{x}, t^{n+1}) = \mathbf{q}^{n+1}(\mathbf{x}) = -k(\mathbf{x})\nabla u^{n+1}(\mathbf{x}) = -k(\mathbf{x})\nabla \mathbf{U}(\mathbf{x})\boldsymbol{\beta}^{n+1} \tag{8.97}$$

Further, the boundary normal heat flux can be written as

$$q_n^{n+1}(\mathbf{x}) = \mathbf{n}^\mathrm{T}\mathbf{q}^{n+1}(\mathbf{x}) = -\mathbf{n}^\mathrm{T}\mathbf{k}(\mathbf{x})\nabla u^{n+1}(\mathbf{x}) = -\mathbf{n}^\mathrm{T}\mathbf{k}(\mathbf{x})\nabla \mathbf{U}(\mathbf{x})\boldsymbol{\beta}^{n+1} \quad (8.98)$$

8.4.1.3 Construction of solving equations

To determine the unknowns $\boldsymbol{\beta}^{n+1}$ in the full temperature solution (8.94), it is required that the approximated temperature solution (8.94) satisfies the governing Eq. (8.91) at M interpolation points in the computational domain, and the boundary conditions (8.92) and (8.93) at N boundary collocations. Thus, substituting Eq. (8.94) into Eq. (8.91) and utilizing M interpolation points in the domain $\varOmega$, one obtains

$$\widetilde{\mathfrak{R}}\big[\mathbf{U}(\mathbf{x}_i)\boldsymbol{\beta}^{n+1}\big] + Q(\mathbf{x}_i) = \widetilde{\mathfrak{R}}[\mathbf{U}(\mathbf{x}_i)]\boldsymbol{\beta}^{n+1} + Q(\mathbf{x}_i) = 0, \ \mathbf{x}_i \in \varOmega, \ i = 1, 2, ..., M$$

$$(8.99)$$

Besides, the satisfaction of the specific boundary conditions (8.92) and (8.93) at N boundary collocations for the temperature solution (8.94) and the normal heat flux solution (8.98) gives

$$\begin{aligned} \mathbf{U}(\mathbf{x}_j)\boldsymbol{\beta}^{n+1} &= \bar{u}\big(\mathbf{x}_j, t^{n+1}\big), & \mathbf{x}_j \in \varGamma_1, \ j = 1, 2, ..., N_1 \\ \big[-\mathbf{n}^\mathrm{T}\mathbf{k}(\mathbf{x}_j)\nabla\mathbf{U}(\mathbf{x}_j)\big]\boldsymbol{\beta}^{n+1} &= \bar{q}\big(\mathbf{x}_j, t^{n+1}\big), & \mathbf{x}_j \in \varGamma_2, \ j = 1, 2, ..., N_2 \end{aligned} \quad (8.100)$$

where N_1 and N_2 are collocations on the boundary $\varGamma_1$ and $\varGamma_2$, respectively, and $N_1 + N_2 = N$.

Combining Eqs. (8.99) and (8.100), we would obtain a linear system of equations:

$$\begin{bmatrix} \widetilde{\mathfrak{R}}[\mathbf{U}(\mathbf{x}_i)] \\ \mathbf{U}(\mathbf{x}_j) \\ -\mathbf{n}^\mathrm{T}\mathbf{k}(\mathbf{x}_j)\nabla\mathbf{U}(\mathbf{x}_j) \end{bmatrix} \boldsymbol{\beta}^{n+1} = \begin{bmatrix} -Q(\mathbf{x}_i) \\ \bar{u}\big(\mathbf{x}_j, t^{n+1}\big) \\ \bar{q}\big(\mathbf{x}_j, t^{n+1}\big) \end{bmatrix} \quad (8.101)$$

or in abbreviated form,

$$\mathbf{A}_{(M+N)\times(M+N)}\boldsymbol{\beta}^{n+1} = \mathbf{f} \quad (8.102)$$

which produces an $(M + N)\times(M + N)$ linear system of equations for the unknown coefficient vector $\boldsymbol{\beta}^{n+1}$. Once these unknown coefficients are determined, the temperature solution u^{n+1} and the heat flux component $\mathbf{q}^{n+1}$ at any field point $\mathbf{x} \in \varOmega$ can be evaluated using Eqs. (8.94) and (8.97).

8.4.2 Results and discussion

To demonstrate the efficiency and accuracy of the proposed meshless method and the selection of RBF and virtual boundary, the transient heat conduction in an isotropic homogeneous square plate is first considered since the corresponding exact solutions can be available for comparison. Then the transient heat conduction in the functionally graded material is discussed.

8.4.2.1 Isotropic homogeneous square plate with sudden temperature jump

Consider a benchmark problem whose geometry is a unit square in which no internal heat source exists, as shown in Fig. 8.26. Zero initial temperature has been assumed, and homogeneous Neumann boundary conditions (insulation) are prescribed on the sides $x = 0$, $y = 0$, and $y = 1$, respectively. The remaining side is subjected to a sudden unit temperature jump, i.e., $u(x = 1, y, t > 0) = 1$, which can be described by the Heaviside step function $H(t)$:

$$u(x = 1, y) = H(t) = \begin{cases} 0.0, & t \le 0 \\ 1.0, & t > 0 \end{cases} \tag{8.103}$$

Obviously, the Heaviside step function $H(t)$ is a discontinuous unit step function, so its derivative is given by

$$\frac{dH(t)}{dt} = \delta(t) \tag{8.104}$$

where $\delta(t)$ is the delta function centered at $t = 0$.

Following the given boundary conditions and the assumption that the thermal conductivity of material is a constant k, the transient heat conduction along the x-direction is defined. Using the method of variable separation, the analytical solution of temperature can be obtained as

$$u(x, t) = 1 - \sum_{i=0}^{\infty} (-1)^i \frac{4}{(2i + 1)\pi} \cos(\mu_i x) \exp(-\mu_i^2 t) \tag{8.105}$$

from which the heat flux components can be obtained as

$$q_x(x, t) = -2k \times \left[\sum_{i=0}^{\infty} (-1)^i \sin(\mu_i x) \exp(-\mu_i^2 t) \right] \tag{8.106}$$

$$q_y(x, t) = 0$$

where

$$\mu_i = \frac{(2i + 1)\pi}{2} \tag{8.107}$$

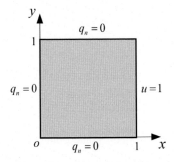

FIGURE 8.26 Schematics of the unit square and the given boundary conditions.

In the computation, thermal diffusivity $a = 1 \text{ m}^2/\text{s}$ and thermal conductivity of materials $k = 1 \text{ W}/(\text{m} \cdot \text{K})$ is assumed. The uniform interpolation scheme is used in the domain. A total of 20 fictitious source points are selected on the virtual boundary and $11 \times 11 = 121$ uniform interpolation points are used for RBF interpolation, unless there is a special statement.

Firstly, both the first- and third-order BRFs, $1 + r$ and $1 + r^3$, which have been widely used in the literature, are used in the calculation, and the corresponding results are compared. The numerical results in Fig. 8.27 show that the use of higher-order RBF interpolation functions does not improve computing accuracy in this transient problem. This phenomenon has also been observed in previous work [19]. Therefore, special care must be taken in using higher-order RBFs since the linear system of equations associated with the computation of the particular solution can easily result in an ill-conditioned number, which is directly linked to the order of the RBFs and the density of the interpolation points. In contrast, the low-order RBF can maintain the necessary accuracy and stability, especially for the time-dependent problem, in which the dominant error is presumably caused by the time-stepping scheme [19,20]. Therefore, the first-order interpolation function $1 + r$, also known as the ad-hoc function, is employed in the following computation.

To assess the effect of the location of the virtual boundary on the accuracy of the proposed algorithm, Fig. 8.28 presents the relative error of temperature versus the similarity ratio at point (0.5, 0.5) with time step $\nabla t = 0.01$ s. The results in Fig. 8.28 show that good computational accuracy and stability are achieved when the similarity ratio is greater than 2, and the optimal value of the similarity ratio is between 2.5 and 5.0. Although the virtual boundary can theoretically be chosen arbitrarily outside of the domain, a distance between

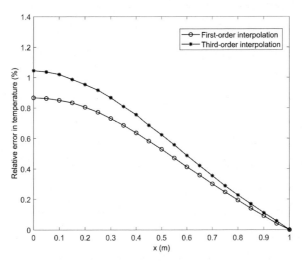

FIGURE 8.27 Effect of the order of RBF at $t = 0.5$ s.

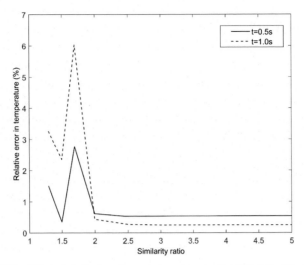

FIGURE 8.28 Effect of similarity ratio on temperature at point (0.5, 0.5) with $\Delta t = 0.01$ s.

the virtual and physical boundaries being either too small or too great will reduce the computational accuracy of the proposed meshless method, due to the singularity of the fundamental solution and the restriction of computer precision, including round-off error [13,21].

Fig. 8.29 shows the percentage error of temperature for two different time steps. It can be seen that the smaller the time step is, the greater the accuracy of the results are obtained. However, more computational cost will inevitably

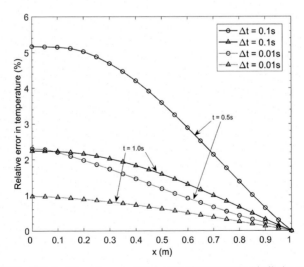

FIGURE 8.29 Effect of time step on relative error of temperature (similarity ratio $= 3.0$).

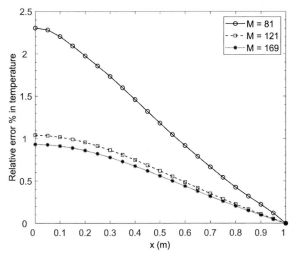

FIGURE 8.30 Effect of the number of interpolation points on the temperature along the *x*-axis for $N = 20$.

be required if a smaller time step is chosen. Additionally, further reduction in the time step does not guarantee the reduction of the relative error, as indicated in the literature [20].

Additionally, the convergence of the present meshless approach for transient heat conduction is investigated as the number of RBF interpolation points M increases. Fig. 8.30 displays that the relative error in temperature along the *x*-axis decreases with the increase of the number of RBF interpolation points M when the number of virtual collocations $N = 20$.

8.4.2.2 Isotropic homogeneous square plate with nonzero initial condition

In this example, the heat dissipation of a square plate with side length L is analyzed. For simplicity, it is still assumed that $k = 1$ W/(m·K) and $a = 1$ m^2/s, as done in the first example. The Dirichlet boundary conditions with zero value of temperature are applied over the four boundary edges, as indicated in Fig. 8.31. Simultaneously, the initial condition for this example is given by

$$u(x, y, 0) = 10 \sin\left(\frac{\pi x}{L}\right) \sin\left(\frac{\pi y}{L}\right) \tag{8.108}$$

Correspondingly, the exact solution of this problem is written as

$$u(x, y, t) = 10 \sin\left(\frac{\pi x}{L}\right) \sin\left(\frac{\pi y}{L}\right) e^{-\frac{2\pi^2 t}{L^2}} \tag{8.109}$$

which will finally lead to a zero-temperature distribution in the domain when $t \to \infty$, as was expected.

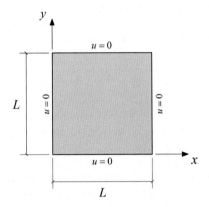

FIGURE 8.31 Schematics of the square domain with zero temperature boundary conditions.

In the computation, the side length of the square is taken as 1.0 m. The number of boundary collocations is 36, and the number of RBF interpolation points is 169. The similarity ratio is 3. Fig. 8.32 depicts the variation of temperature at the central point (0.5 m, 0.5 m) in terms of time for various time increments. It is found that the smaller time increment can lead to the better accuracy in the present meshless method. However, the smaller time increment causes more numerical iterations and computation time. Besides, in Figs. 8.33 and 8.34, the variations of the temperature u and the heat flux component q_x are plotted along the x-axis at different time instances. It is clearly seen that both the temperature and the heat flux component decrease as

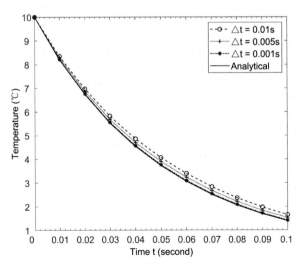

FIGURE 8.32 Variation of temperature at the central point (0.5, 0.5) as the time increment changes.

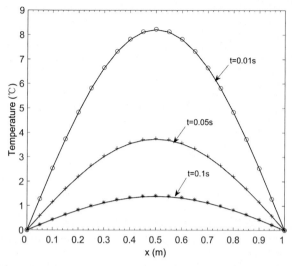

FIGURE 8.33   Variations of temperature along the axis $y = 0$ at various time instances.

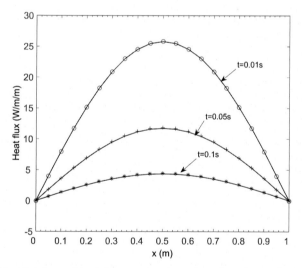

FIGURE 8.34   Variations of heat flux component q_x along the axis $y = 0$ at various time instances.

the time increases. This can be attributed to that the influence of the initial temperature becomes weak when the time increases, and the temperature distribution finally tends to zero value, which is consistent with the given temperature boundary conditions. Besides, the solid lines in the figures represent the analytical solutions. A good agreement between the numerical results and the analytical solutions is observed.

8.4.2.3 Isotropic homogeneous square plate with cone-shaped solution

In the third, the transient heat conduction with a cone-shaped solution is considered in a unit square plate. The material properties are $k = 1$ W/m·°C and $\rho c = 1$ J/m³·°C. The right-handed term Q in Eq. (8.81), the Dirichlet boundary conditions prescribed on all sides, and the initial condition are selected, so the exact solution is as follows [22]:

$$u(x, y, t) = 0.8 \exp\left\{-80\left[(x-r)^2 + (y-s)^2\right]\right\} \tag{8.110}$$

where

$$r(t) = \frac{1}{4}(2 + \sin \pi t), \quad s(t) = \frac{1}{4}(2 + \cos \pi t) \tag{8.111}$$

The exact solution is a cone that is initially centered at point (1/2, 3/4) and then rotates around the center of the domain in a clockwise direction for $t > 0$. The peak of the cone is 0.8.

This problem is solved for $0 < t < 2$ (one period) using time step $\Delta t = 0.01$ s. The square domain is discretized by 13 by 13 RBF collocations uniformly gridded and 36 virtual source points. The similarity ratio is 3. Fig. 8.35 shows the approximate solutions for the present meshless method, from which it is observed that the numerical results are accurate and stable. The maximum of the numerical solution is 0.7896.

8.4.2.3.1 Isotropic functionally graded finite strip

Consider a functionally graded finite strip with a unidirectional variation of thermal conductivity [23]. Zero initial temperature is assumed. The same exponential spatial variation for the thermal conductivity and the diffusivity is assumed as

$$k(\mathbf{x}) = k_0 e^{\eta x} \tag{8.112}$$

$$a = \frac{k}{\rho c} = a_0 e^{\eta x} \tag{8.113}$$

where $k_0 = 17$ W/m·°C and $a_0 = 0.17 \times 10^{-4}$ m²/s.

Two different exponential parameters $\eta = 0.2$ and 0.5 cm^{-1} are used in the numerical calculation. On the sides parallel to the y-axis, two different temperatures are prescribed. One side is kept at zero temperature and the other has the Heaviside function of time, i.e., $u = T \cdot H(t)$ with $T = 1$ °C. On the lateral sides of the strip the heat flux vanishes. In the numerical calculation, a square with side length $L = 0.04$ m is considered (see Fig. 8.36).

The special case with an exponential parameter $\eta = 0$ is considered first. In this case the analytical solution is written as

$$u(x, t) = T\frac{x}{L} + \frac{2}{\pi}\sum_{n=1}^{\infty}\frac{T\cos n\pi}{n}\sin\frac{n\pi x}{L}\exp\left(-\frac{an^2\pi^2 t}{L^2}\right) \tag{8.114}$$

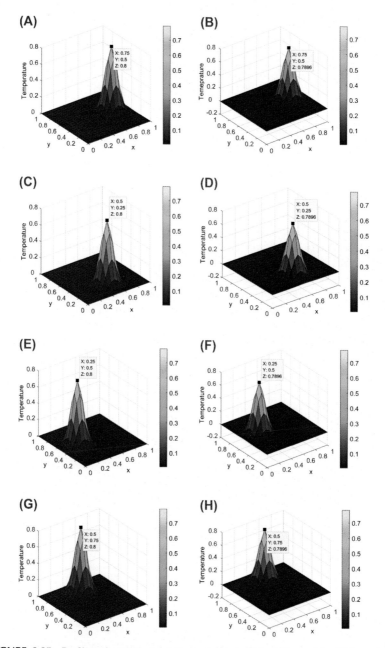

FIGURE 8.35 Profiles of analytical solutions at (A) $t = 0.5$ s, (C) $t = 1.0$ s, (E) $t = 1.5$ s, (G) $t = 2.0$ s and numerical results at (B) $t = 0.5$ s, (D) $t = 1.0$ s, (F) $t = 1.5$ s, (H) $t = 2.0$ s.

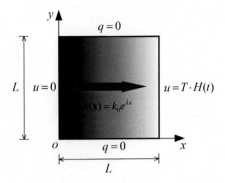

FIGURE 8.36 Geometry of a functionally graded finite square strip and boundary conditions.

which can be used to check the accuracy of the present numerical method. Numerical results are obtained using 36 fictitious source points, 169 interpolation points, similarity ratio = 3.0, and time step $\Delta t = 1$ s. The following computation is carried out using the first-order interpolation function $1 + r$ only. Fig. 8.37 shows the temperature field at the three points ($x = 0.01$, 0.02, and 0.03 m). A good agreement between numerical and analytical results is observed from Fig. 8.37.

The preceding discussion concerns heat conduction in homogeneous materials only, for which the analytical solutions are available. To illustrate the application of the proposed algorithm to the functionally graded materials (FGM), consider now the FGM with $\eta = 0.2$ and 0.5 cm^{-1}, respectively. The variation of temperature at position $x = 0.02$ m with time for three η values is

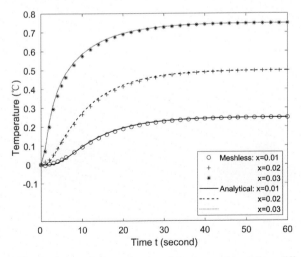

FIGURE 8.37 Time variation of temperature in a finite square strip at three different positions with $\eta = 0$.

presented in Fig. 8.38. Fig. 8.39 shows the distribution of temperature along the x-axis at $t = 30$ s. As was expected, it is found from Fig. 8.38 that the temperature increases along with an increase in η values (or equivalently in thermal conductivity), and the temperature approaches a steady state when $t > 20$ s. It is observed from Fig. 8.39 that the temperature increases along with an increase in η values again.

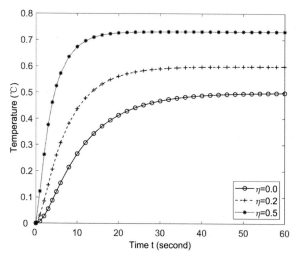

FIGURE 8.38 Time variation of temperature at the position $x = 0.02$ m of a functionally graded finite square strip.

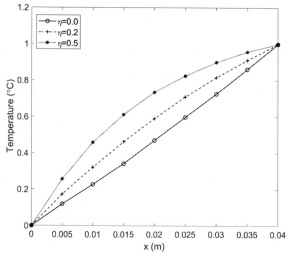

FIGURE 8.39 Distribution of temperature at $t = 30$ s along the x-axis of a functionally graded finite square strip.

For the final steady state, an analytical solution can be obtained as

$$u(x) = T \frac{e^{-\eta x} - 1}{e^{-\eta a} - 1} \qquad (8.115)$$

which has a limit value when $\eta \to 0$:

$$\lim_{\eta \to 0} u(x) = T \frac{x}{a} \qquad (8.116)$$

Analytical and numerical results obtained at the time $t = 70$ s corresponding to stationary or static loading conditions are presented in Fig. 8.40. The numerical results are in good agreement with the analytical results for the case of steady state.

In addition, the same problem is analyzed by Sladek et al. [23] using the meshless local boundary integral equation method (LBIEM). Here, we make a comparison to demonstrate the proposed method. From Table 8.1 we can see

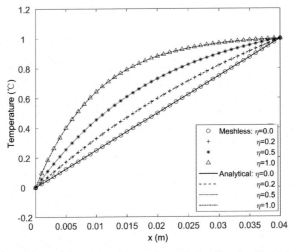

FIGURE 8.40   Distribution of temperature along the x-axis for a functionally graded finite square strip under steady-state loading conditions.

TABLE 8.1 Comparison of LBIEM and the proposed method at $\eta = 0.5$ cm^{-1} and $x = 0.01$ m.

	$t = 10$ s	$t = 20$ s	$t = 30$ s	$t = 40$ s	$t = 50$ s	$t = 60$ s	$t = $ infinite
LBIEM	0.1871	0.3281	0.3800	0.3986	0.4019	0.4053	0.4581
Present	0.3915	0.4497	0.4546	0.4550	0.4551	0.4551	0.4551
Exact	/	/	/	/	/	/	0.4551

that the proposed method provided a little bit larger values of temperature at same instance than those from the LBIEM. After $t = 50$ s a relatively steady state is reached. It should be noted that the results in Figs. 4 and 5 in the literature [23] are slightly different than the numerical results in this study. The different treatment of time domain is probably the main reason for this difference. In the LBIEM, the Laplace transformation technology is used instead of the time-stepping scheme employed in the proposed method. However, to the steady-state temperature field at $x = 0.01$ m the results are similar between the two methods.

8.5 Remarks

In this chapter, the steady-state and transient heat conduction problems are solved by the meshless method. In the present meshless method, the temperature solution for the steady-state heat conduction problems in heterogeneous materials is approximated as the general solution of Laplace equation by means of the analog equation method. Then, the boundary collocations and the RBF interpolation points are respectively used to construct the expressions of the corresponding homogeneous and particular solutions. Finally, the satisfaction of the original governing equation and the given boundary conditions can be used to determine all unknowns in the expression of general temperature solution. In addition, when the transient heat conduction problems are discretized in the time domain using the time-stepping scheme, they can be converted to a series of partial differential equations at different time instances, which can be solved by the present meshless method, as done in the case of steady-state heat conduction. Numerical results reveal that the accuracy of the present meshless method behaves well, so it is concluded that the present meshless method can be employed for the simulation of steady-state and transient heat conduction problems in general material definitions. Moreover, with the present solving strategy, the nonlinear heat conduction problems can also be solved conveniently, in which the thermal conductivity of material is a function of temperature.

References

[1] Y.P. Chang, C.S. Kang, D.J. Chen, The use of fundamental green's functions for the solution of problems of heat conduction in anisotropic media, International Journal of Heat and Mass Transfer 16 (1973) 1905–1918.

[2] B. Jin, L. Marin, The method of fundamental solutions for inverse source problems associated with the steady-state heat conduction, International Journal for Numerical Methods in Engineering 69 (2007) 1570–1589.

[3] H. Wang, Q.H. Qin, Y.L. Kang, A new meshless method for steady-state heat conduction problems in anisotropic and inhomogeneous media, Archive of Applied Mechanics 74 (2005) 563–579.

[4]   B.T. Johansson, D. Lesnic, A method of fundamental solutions for transient heat conduction, Engineering Analysis with Boundary Elements 32 (2008) 697–703.

[5]   B.T. Johansson, Properties of a method of fundamental solutions for the parabolic heat equation, Applied Mathematics Letters 65 (2017) 83–89.

[6]   H. Wang, Q.H. Qin, Y.L. Kang, A meshless model for transient heat conduction in functionally graded materials, Computational Mechanics 38 (2006) 51–60.

[7]   Z.W. Zhang, H. Wang, Q.H. Qin, Method of fundamental solutions for nonlinear skin bioheat model, Journal of Mechanics in Medicine and Biology 14 (2014) 1450060.

[8]   Z.W. Zhang, H. Wang, Q.H. Qin, Meshless method with operator splitting technique for transient nonlinear bioheat transfer in two-dimensional skin tissues, International Journal of Molecular Sciences 16 (2015) 2001–2019.

[9]   H. Wang, Q.H. Qin, A meshless method for generalized linear or nonlinear Poisson-type problems, Engineering Analysis with Boundary Elements 30 (2006) 515–521.

[10] L. Marin, D. Lesnic, The method of fundamental solutions for nonlinear functionally graded materials, International Journal of Solids and Structures 44 (2007) 6878–6890.

[11] J.H. Lienhard IV, J.H. Lienhard V, A Heat Transfer Textbook, Phlogiston Press, Cambridge, Massachusetts, U.S.A., 2005.

[12] J.T. Katsikadelis, The Boundary Element Method for Engineers and Scientists: Theory and Applications, Elsevier, 2016.

[13] P. Mitic, Y.F. Rashed, Convergence and stability of the method of meshless fundamental solutions using an array of randomly distributed sources, Engineering Analysis with Boundary Elements 28 (2004) 143–153.

[14] P.W. Partridge, C.A. Brebbia, L.C. Wrobel, The Dual Reciprocity Boundary Element Method, Springer, 1991.

[15] T. Yamada, L.C. Wrobel, H. Power, On the convergence of the dual reciprocity boundary element method, Engineering Analysis with Boundary Elements 13 (1994) 291–298.

[16] A.J. Kassab, E. Divo, A generalized boundary integral equation for isotropic heat conduction with spatially varying thermal conductivity, Engineering Analysis with Boundary Elements 18 (1996) 273–286.

[17] E. Divo, A.J. Kassab, Generalized boundary integral equation for heat conduction in non-homogeneous media: recent developments on the sifting property, Engineering Analysis with Boundary Elements 22 (1998) 221–234.

[18] N.S. Mera, L. Elliott, D.B. Ingham, D. Lesnic, A comparison of boundary element method formulations for steady state anisotropic heat conduction problems, Engineering Analysis with Boundary Elements 25 (2001) 115–128.

[19] M.S. Ingber, C.S. Chen, J.A. Tanski, A mesh free approach using radial basis functions and parallel domain decomposition for solving three-dimensional diffusion equations, International Journal for Numerical Methods in Engineering 60 (2004) 2183–2201.

[20] R. Schaback, Error estimates and condition numbers for radial basis function interpolation, Advances in Computational Mathematics 3 (1995) 251–264.

[21] H.C. Sun, L.Z. Zhang, Q. Xu, Y.M. Zhang, Nonsingularity Boundary Element Methods (In Chinese), Dalian University of Technology Press, Dalian, 1999.

[22] J. Li, Y.C. Hon, C.S. Chen, Numerical comparisons of two meshless methods using radial basis functions, Engineering Analysis with Boundary Elements 26 (2002) 205–225.

[23] J. Sladek, V. Sladek, C.Z. Zhang, Transient heat conduction analysis in functionally graded materials by the meshless local boundary integral equation method, Computational Materials Science 28 (2003) 494–504.

Appendix A

Derivatives of functions in terms of radial variable *r*

In the textbook, the radial basis function frequently employed is in terms of radial variable r only, so this requires us to express derivatives of such function with respect to the Cartesian coordinates x and y, which are provided below for reference.

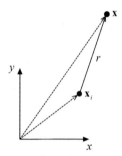

FIGURE A.1 Definition of Euclidean distance r.

For a two-dimensional problem, the radial distance can be defined as follows (see Fig. A.1):

$$r = |\mathbf{x} - \mathbf{x}_i| = \sqrt{r_1^2 + r_2^2} \tag{A.1}$$

where

$$r_1 = x - x_i, \quad r_2 = y - y_i \tag{A.2}$$

and $\mathbf{x} = (x, y)$, $\mathbf{x}_i = (x_i, y_i)$

Obviously, for any smooth function $f(r)$, the familiar tree diagram shown in Fig. A.2 can be established.

FIGURE A.2 Familiar tree diagram.

Firstly, the spatial derivative of r can be written as

$$\frac{\partial r}{\partial x} = \frac{r_1}{r} \tag{A.3}$$

$$\frac{\partial r}{\partial y} = \frac{r_2}{r} \tag{A.4}$$

$$\frac{\partial^2 r}{\partial x^2} = \frac{r_2^2}{r^3} \tag{A.5}$$

$$\frac{\partial^2 r}{\partial y^2} = \frac{r_1^2}{r^3} \tag{A.6}$$

$$\frac{\partial^2 r}{\partial x \partial y} = \frac{r_1 r_2}{r^3} \tag{A.7}$$

$$\frac{\partial^3 r}{\partial x^3} = -\frac{3 r_1 r_2^2}{r^5} \tag{A.8}$$

$$\frac{\partial^3 r}{\partial y^3} = -\frac{3 r_2 r_1^2}{r^5} \tag{A.9}$$

$$\frac{\partial^3 r}{\partial x^2 \partial y} = -\frac{3 r_2^3}{r^5} + \frac{2 r_2}{r^3} \tag{A.10}$$

$$\frac{\partial^3 r}{\partial x \partial y^2} = -\frac{3 r_1^3}{r^5} + \frac{2 r_1}{r^3} \tag{A.11}$$

$$\frac{\partial^4 r}{\partial x^4} = -\frac{15 r_1^4}{r^7} + \frac{18 r_1^2}{r^5} - \frac{3}{r^3} \tag{A.12}$$

$$\frac{\partial^4 r}{\partial y^4} = -\frac{15 r_2^4}{r^7} + \frac{18 r_2^2}{r^5} - \frac{3}{r^3} \tag{A.13}$$

$$\frac{\partial^4 r}{\partial x^2 \partial y^2} = -\frac{15 r_1^2 r_2^2}{r^7} + \frac{2}{r^3} \tag{A.14}$$

from which we have some useful conclusions:

$$\left(\frac{\partial r}{\partial x}\right)^2 + \left(\frac{\partial r}{\partial y}\right)^2 = \left(\frac{r_1}{r}\right)^2 + \left(\frac{r_2}{r}\right)^2 \equiv 1 \tag{A.15}$$

$$\frac{\partial^2 r}{\partial x^2} + \frac{\partial^2 r}{\partial y^2} = \frac{r_2^2}{r^3} + \frac{r_1^2}{r^3} = \frac{1}{r} \tag{A.16}$$

Then, following the chain rule of differentiation, we have

$$\frac{\partial f}{\partial x} = \frac{df}{dr}\frac{\partial r}{\partial x} \tag{A.17}$$

$$\frac{\partial f}{\partial y} = \frac{df}{dr}\frac{\partial r}{\partial y} \tag{A.18}$$

$$\frac{\partial^2 f}{\partial x^2} = \frac{d^2 f}{dr^2}\left(\frac{\partial r}{\partial x}\right)^2 + \frac{df}{dr}\frac{\partial^2 r}{\partial x^2} \tag{A.19}$$

$$\frac{\partial^2 f}{\partial y^2} = \frac{d^2 f}{dr^2}\left(\frac{\partial r}{\partial y}\right)^2 + \frac{df}{dr}\frac{\partial^2 r}{\partial y^2} \tag{A.20}$$

$$\frac{\partial^2 f}{\partial x \partial y} = \frac{d^2 f}{dr^2}\frac{\partial r}{\partial x}\frac{\partial r}{\partial y} + \frac{df}{dr}\frac{\partial^2 r}{\partial x \partial y} \tag{A.21}$$

$$\frac{\partial^3 f}{\partial x^3} = \frac{d^3 f}{dr^3}\left(\frac{\partial r}{\partial x}\right)^3 + 3\frac{d^2 f}{dr^2}\frac{\partial r}{\partial x}\frac{\partial^2 r}{\partial x^2} + \frac{df}{dr}\frac{\partial^3 r}{\partial x^3} \tag{A.22}$$

$$\frac{\partial^3 f}{\partial y^3} = \frac{d^3 f}{dr^3}\left(\frac{\partial r}{\partial y}\right)^3 + 3\frac{d^2 f}{dr^2}\frac{\partial r}{\partial y}\frac{\partial^2 r}{\partial y^2} + \frac{df}{dr}\frac{\partial^3 r}{\partial y^3} \tag{A.23}$$

$$\frac{\partial^3 f}{\partial x^2 \partial y} = \frac{d^3 f}{dr^3}\left(\frac{\partial r}{\partial x}\right)^2\frac{\partial r}{\partial y} + \frac{d^2 f}{dr^2}\left(2\frac{\partial r}{\partial x}\frac{\partial^2 r}{\partial x \partial y} + \frac{\partial r}{\partial y}\frac{\partial^2 r}{\partial x^2}\right) + \frac{df}{dr}\frac{\partial^3 r}{\partial x^2 \partial y} \tag{A.24}$$

$$\frac{\partial^3 f}{\partial x \partial y^2} = \frac{d^3 f}{dr^3}\left(\frac{\partial r}{\partial y}\right)^2\frac{\partial r}{\partial x} + \frac{d^2 f}{dr^2}\left(2\frac{\partial r}{\partial y}\frac{\partial^2 r}{\partial x \partial y} + \frac{\partial r}{\partial x}\frac{\partial^2 r}{\partial y^2}\right) + \frac{df}{dr}\frac{\partial^3 r}{\partial x \partial y^2} \tag{A.25}$$

$$\frac{\partial^4 f}{\partial x^4} = \frac{d^4 f}{dr^4}\left(\frac{\partial r}{\partial x}\right)^4 + 6\frac{d^3 f}{dr^3}\left(\frac{\partial r}{\partial x}\right)^2\frac{\partial^2 r}{\partial x^2} + \frac{d^2 f}{dr^2}\left[3\left(\frac{\partial r}{\partial x}\right)^2 + 4\frac{\partial r}{\partial x}\frac{\partial^3 r}{\partial x^3}\right] + \frac{df}{dr}\frac{\partial^4 r}{\partial x^4} \tag{A.26}$$

$$\frac{\partial^4 f}{\partial y^4} = \frac{d^4 f}{dr^4}\left(\frac{\partial r}{\partial y}\right)^4 + 6\frac{d^3 f}{dr^3}\left(\frac{\partial r}{\partial y}\right)^2\frac{\partial^2 r}{\partial y^2} + \frac{d^2 f}{dr^2}\left[3\left(\frac{\partial r}{\partial y}\right)^2 + 4\frac{\partial r}{\partial y}\frac{\partial^3 r}{\partial y^3}\right] + \frac{df}{dr}\frac{\partial^4 r}{\partial y^4}$$

$$(A.27)$$

$$\frac{\partial^4 f}{\partial x^2 \partial y^2} = \frac{d^4 f}{dr^4}\left(\frac{\partial r}{\partial x}\right)^2\left(\frac{\partial r}{\partial y}\right)^2 + \frac{d^3 f}{dr^3}\left[\left(\frac{\partial r}{\partial y}\right)^2\frac{\partial^2 r}{\partial x^2} + \left(\frac{\partial r}{\partial x}\right)^2\frac{\partial^2 r}{\partial y^2} + 4\frac{\partial r}{\partial x}\frac{\partial r}{\partial y}\frac{\partial^2 r}{\partial x \partial y}\right]$$

$$+ \frac{d^2 f}{dr^2}\left[2\left(\frac{\partial^2 r}{\partial x \partial y}\right)^2 + 2\frac{\partial r}{\partial x}\frac{\partial^3 r}{\partial x \partial y^2} + 2\frac{\partial r}{\partial y}\frac{\partial^3 r}{\partial x^2 \partial y} + \frac{\partial^2 r}{\partial x^2}\frac{\partial^2 r}{\partial y^2}\right] + \frac{df}{dr}\frac{\partial^4 r}{\partial x^2 \partial y^2}$$

$$(A.28)$$

Appendix B

Transformations

In the practical analysis, some quantities like the displacements, stresses, or electric displacement might be expressed in terms of the polar coordinates, which are generally different from the Cartesian coordinates defined for the problem to describe the physical behavior of a continuum body. Therefore, it is sometimes convenient to introduce a rule that can transform such field variables from one coordinate system to another.

B.1 Coordinate transformation

In the two-dimensional Cartesian coordinate system, the two coordinates are perpendicular to one another with the same unit length on both axes. While in the two-dimensional polar coordinate system, each point is determined by an angle relative to the zero axis and a distance to the origin of system. Fig. B.1 shows the classic Cartesian and polar coordinate systems, from which it is observed that each point is determined by coordinates (x, y) in the Cartesian system and by polar coordinates (r, θ) in the polar system. Also, it is easily found from Fig. B.1 that the connection between Cartesian and polar coordinates can be written as

$$\begin{cases} x = r \cos \theta \\ y = r \sin \theta \end{cases} \tag{B.1}$$

or

$$\begin{cases} r = \sqrt{x^2 + y^2} \\ \theta = \arctan \dfrac{y}{x} \end{cases} \tag{B.2}$$

In some cases, the introduction of polar coordinates might make the expression present in a more concise form. For example, the expression (see Fig. B.2)

$$\arctan \frac{y}{x} = \sqrt{x^2 + y^2} \tag{B.3}$$

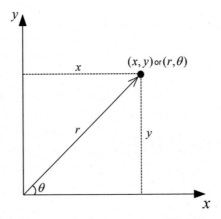

FIGURE B.1 The connection between polar and Cartesian coordinates.

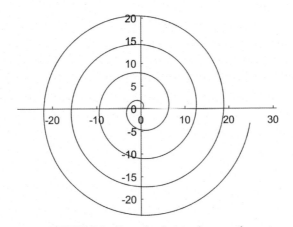

FIGURE B.2 Example of expression $r = \theta$.

presented in Cartesian coordinates can be rewritten as in polar coordinates

$$r = \theta \tag{B.4}$$

B.2 Vector transformation

An arbitrary vector **u** may be written in the Cartesian coordinates and the polar coordinates as

$$\mathbf{u} = u_x\mathbf{i} + u_y\mathbf{j} = u_r\mathbf{e}_r + u_\theta\mathbf{e}_\theta \tag{B.5}$$

where **i**, **j** are unit direction vectors along the x and y axes in the Cartesian coordinate system, while $\mathbf{e}_r$ and $\mathbf{e}_\theta$ are unit direction vectors along the r and θ axes in the polar coordinate system, as shown in Fig. B.3.

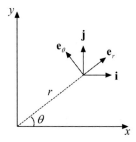

FIGURE B.3 Connection of unit direction vectors in the Cartesian and polar coordinates.

By means of vector decomposition, we can take

$$\begin{aligned}
\mathbf{e}_r &= \cos\theta\mathbf{i} + \sin\theta\mathbf{j} \\
\mathbf{e}_\theta &= -\sin\theta\mathbf{i} + \cos\theta\mathbf{j}
\end{aligned} \tag{B.6}$$

or

$$\begin{aligned}
\mathbf{i} &= \cos\theta\mathbf{e}_r - \sin\theta\mathbf{e}_\theta \\
\mathbf{j} &= \sin\theta\mathbf{e}_r + \cos\theta\mathbf{e}_\theta
\end{aligned} \tag{B.7}$$

Substituting Eq. (B.6) into Eq. (B.5) yields

$$\begin{Bmatrix} u_x \\ u_y \end{Bmatrix} = \begin{bmatrix} \cos\theta & -\sin\theta \\ \sin\theta & \cos\theta \end{bmatrix} \begin{Bmatrix} u_r \\ u_\theta \end{Bmatrix} \tag{B.8}$$

Similarly, we have the inverse transformation by substituting Eq. (B.7) into Eq. (B.5):

$$\begin{Bmatrix} u_r \\ u_\theta \end{Bmatrix} = \begin{bmatrix} \cos\theta & \sin\theta \\ -\sin\theta & \cos\theta \end{bmatrix} \begin{Bmatrix} u_x \\ u_y \end{Bmatrix} \tag{B.9}$$

B.3 Stress transformation

Different from the vector, the conversion of stress tensor consisting of three components in two-dimensional space, as given in Fig. B.4, cannot be derived from the Cartesian coordinates to the polar coordinates by the vector decomposition. In practice, by considering the equilibrium of microelements in Fig. B.5, we can obtain the following relation:

$$\begin{Bmatrix} \sigma_r \\ \sigma_\theta \\ \tau_{r\theta} \end{Bmatrix} = \begin{bmatrix} \cos^2\theta & \sin^2\theta & \sin 2\theta \\ \sin^2\theta & \cos^2\theta & -\sin 2\theta \\ -\dfrac{\sin 2\theta}{2} & \dfrac{\sin 2\theta}{2} & \cos 2\theta \end{bmatrix} \begin{Bmatrix} \sigma_x \\ \sigma_y \\ \tau_{xy} \end{Bmatrix} \tag{B.10}$$

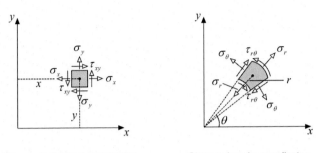

Stresses in Carteian coordinates   Stresses in polar coordinates

FIGURE B.4 Stress components in the Cartesian and polar coordinates.

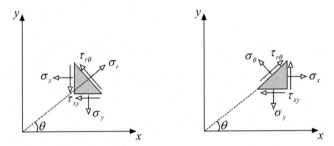

FIGURE B.5 Equilibrium of microelements.

Conversely, the stress transformation from the polar coordinate system to the Cartesian coordinate system can be derived by the inverse of Eq. (B.10):

$$
\left\{ \begin{array}{c} \sigma_x \\ \sigma_y \\ \tau_{xy} \end{array} \right\} =
\begin{bmatrix}
\cos^2 \theta & \sin^2 \theta & -\sin 2\theta \\
\sin^2 \theta & \cos^2 \theta & \sin 2\theta \\
\dfrac{\sin 2\theta}{2} & -\dfrac{\sin 2\theta}{2} & \cos 2\theta
\end{bmatrix}
\left\{ \begin{array}{c} \sigma_r \\ \sigma_\theta \\ \tau_{r\theta} \end{array} \right\}
\tag{B.11}
$$

Appendix C

Derivatives of approximated particular solutions in inhomogeneous plane elasticity

In the meshless analysis of two-dimensional elastic problems, some quantities are related to the first and second derivatives of the approximated particular solutions, whose expressions are dependent on the given radial basis functions. Therefore, it is necessary to provide the detailed formulations of the first and second derivatives of the approximated particular solutions, for the sake of completeness.

C.1 Power spline (PS) function

$$\Phi_{li} = -\frac{1}{2\mu(1-v)} \frac{1}{(2n+1)^2(2n+3)} r^{2n+1} \left(A_1\delta_{li} + A_2 r_{,l}r_{,i}\right)$$

$$\Phi_{li,j} = -\frac{1}{2\mu(1-v)} \frac{1}{(2n+1)^2(2n+3)} r^{2n} \left[B_1\delta_{li}r_{,j} + B_2\left(\delta_{lj}r_{,i} + \delta_{ij}r_{,l}\right) + B_3 r_{,i}r_{,j}r_{,l}\right]$$

$$\Phi_{lk,ki} = -\frac{1}{2\mu(1-v)} \frac{1}{(2n+1)^2(2n+3)} r^{2n-1} B_4 \left[\delta_{li} + (2n-1)r_{,l}r_{,i}\right]$$

$$\Phi_{li,kk} = -\frac{1}{2\mu(1-v)} \frac{1}{(2n+1)^2(2n+3)} r^{2n-1} \left(C_1\delta_{li} + C_2 r_{,l}r_{,i}\right)$$

where

$$A_1 = (4n+5) - 2v(2n+3)$$

$$A_2 = -(2n + 1)$$

$$B_1 = A_1(2n + 1)$$

$$B_2 = A_2$$

$$B_3 = A_2(2n - 1)$$

$$B_4 = B_1 + 3B_2 + B_3$$

$$C_1 = 2B_2 + B_1(2n + 1)$$

$$C_2 = 2B_2(2n - 1) + B_3(2n + 1)$$

C.2 Thin plate spline (TPS) function

$$\Phi_{li} = -\frac{1}{32\widehat{\mu}\left(1 - \widehat{v}\right)} \frac{r^{2n+2}}{(n+1)^3(n+2)^2}\left(A_1\delta_{il} + A_2 r_{,i}r_{,l}\right)$$

$$\Phi_{li,j} = -\frac{1}{32\widehat{\mu}\left(1 - \widehat{v}\right)} \frac{r^{2n+1}}{(n+1)^3(n+2)^2}$$
$$\left[B_1 r_{,i}r_{,j}r_{,l} + B_2\delta_{il}r_{,j} + B_3\left(\delta_{ij}r_{,l} + \delta_{lj}r_{,i}\right)\right]$$

$$\Phi_{lk,ki} = -\frac{1}{32\widehat{\mu}\left(1 - \widehat{v}\right)} \frac{r^{2n}}{(n+1)^3(n+2)^2}\left(C_1 r_{,i}r_{,l} + B_4\delta_{li}\right)$$

$$\Phi_{li,kk} = -\frac{1}{32\widehat{\mu}\left(1 - \widehat{v}\right)} \frac{r^{2n}}{(n+1)^3(n+2)^2}\left(C_2 r_{,l}r_{,i} + C_3\delta_{il}\right)$$

where

$$A_1 = -\left(8n^2 + 29n + 27\right) + 8\widehat{v}(n+2)^2$$
$$+ 2(n+1)(n+2)\left[4n + 7 - 4\widehat{v}(n+2)\right]\ln r$$

$$A_2 = 2(n+1)[(2n+3) - 2(n+1)(n+2)\ln r]$$

$$B_1 = 2nA_2 - 4(n+1)^2(n+2)$$

$$B_2 = 2(n+1)\left\{A_1 + (n+2)\left[4n+7-4\widehat{v}(n+2)\right]\right\}$$

$$B_3 = A_2$$

$$B_4 = B_1 + B_2 + 3B_3$$

$$C_1 = 2nB_4 + 8(n+1)^2(n+2)^2\left(1-2\widehat{v}\right)$$

$$C_2 = 2(n+1)B_1 + 4nB_3 - 8(n+1)^3(n+2)$$

$$C_3 = 2(n+1)B_2 + 2B_3 - 4(n+1)^2(n+2)\left[-4n-7+4\widehat{v}(n+2)\right]$$

Index

289

Printed in the United States
By Bookmasters